Birkhäuser

Progress in Mathematical Physics

Volume 78

More information about this series at http://www.springer.com/series/4813

Bertrand Duplantier • Vincent Rivasseau
Editors

Information Theory

Poincaré Seminar 2018

Birkhäuser

Editors
Bertrand Duplantier
Institut de Physique Théorique
Université Paris-Saclay, CNRS, CEA
Gif-sur-Yvette Cedex, Essonne, France

Vincent Rivasseau
Laboratoire de Physique Théorique
Université Paris-Saclay, CNRS,
Université Paris-Sud
Orsay, Essonne, France

ISSN 1544-9998 ISSN 2197-1846 (electronic)
Progress in Mathematical Physics
ISBN 978-3-030-81479-3 ISBN 978-3-030-81480-9 (eBook)
https://doi.org/10.1007/978-3-030-81480-9

This book is published under the imprint Birkhäuser, www.birkhauser-science.com by the registered company Springer Nature Switzerland AG
The registered company address is: Gewerbestrasse 11, 6330 Cham, Switzerland

Contents

Foreword

This book is the eighteenth in a series of Proceedings for the *Séminaire Poincaré*, which is directed towards a broad audience of physicists, mathematicians, philosophers and historians of science.

The goal of the Poincaré Seminar is to provide up-to-date information about general topics of great interest in physics. Both the theoretical and experimental aspects of the topic are covered, generally with some historical background. Inspired by the *Nicolas Bourbaki Seminar* in mathematics, hence nicknamed *"Bourbaphy"*, the Poincaré Seminar is held once or twice a year at the Institut Henri Poincaré in Paris, with written contributions prepared in advance. Particular care is devoted to the pedagogical nature of the presentations, so that they may be accessible to a large audience of scientists.

This new volume of the Poincaré Seminar Series, **Information Theory**, corresponds to the twenty-third such seminar, held on November 17th, 2018, at Institut Henri Poincaré in Paris. Its aim is to provide a thorough description of information theory and some of its most active areas, in particular, its relation to thermodynamics at the nanoscale and the Maxwell demon, and the emergence of quantum computation and of its counterpart, quantum verification.

The first article, entitled *Thermodynamics and Information Theory*, by the theoretical physicists KIRONE MALLICK and BERTRAND DUPLANTIER from the Institut de Physique Théorique at Université Paris-Saclay, begins with a review of the laws of classical thermodynamics and statistical physics with an emphasis on the dynamical properties of equilibrium as exemplified by Brownian fluctuations, a universal characteristic of all complex systems at finite temperature. Then it presents recent advances on systems far from equilibrium, which lie beyond the realm of traditional thermodynamics by continuously exchanging matter, energy, or information with their surroundings. Two major contemporary results are the Gallavotti-Cohen-Evans-Morriss *fluctuation theorem* that characterizes statistical properties of entropy production rate, and Jarzynski's *work relation* that extends the XIXth century maximal work inequality to a precise identity, with far-reaching consequences, particularly in studies of active matter. The fluctuation theorem and the work relations are fingerprints at the macroscopic level of the fundamental microscopic time-reversal invariance, satisfied by most systems in condensed matter physics. The last sections of this chapter relate thermodynamics with information theory, starting from Shannon's definition, and proceeding through a description of Maxwell's paradoxical demon, which acquires and exploits information to violate the Second Law of Thermodynamics. Various attempts to exorcise this demon are sketched, from Leó Szilárd's seminal work, Brillouin's brilliant and bold assumption that Shannon's information contributes to the total entropy to preserve the

Second Law's balance sheet, to Landauer's *principle of information erasure.* This introductory chapter paves the way to the more detailed discussions presented in the next articles.

The second article, entitled *This is IT: A Primer on Shannon's Entropy and Information*, is written by OLIVIER RIOUL, a leading researcher in information theory (IT) and signal processing from Télécom Paris of the Institut Polytechnique de Paris. As an enthusiast of Shannon's life and work (see the wonderful movie by Mark A. Levinson, `http://www.documentarymania.com/player.php?title=The%20Bit%20Player`), he offers an in-depth introductory text from both historical and mathematical viewpoints, which lets the elegance and beauty of the subject matter shine through many aspects of the concepts of information and entropy. Thanks to its thirty-seven concise and mostly independent sections, we can appreciate Shannon's 1948 seminal work, *A Mathematical Theory of Communication* (see `http://bibnum.education.fr/sites/default/files/174-article.pdf`), and its mathematical connections to the manifold notions of entropy. The author carefully defines all basic notions of discrete and continuous entropies and entropy power, explains a fundamental information inequality originally due to Gibbs in relation to the MaxEnt (maximal entropy) principle, on which Statistical Mechanics can be entirely founded as advocated by Jaynes, and goes all the way through the notion of relative entropy (or divergence), Chernov, Fisher, and Kolmogorov informations, and, of course, Shannon's mutual information. Shannon's celebrated first and second coding theorems are all stated with sketchs of proofs. This delightful and mostly complete expository text ends with what is perhaps the most difficult inequality due to Shannon, the so-called *entropy power inequality* (EPI), which states that the entropy power of the sum of independent random variables is no less than the sum of their individual entropy powers. Initially used by Shannon for evaluating the capacity of non-Gaussian channels, the EPI now finds multiple applications in IT and mathematics in general. While Shannon's original proof was incomplete and was corrected ten years later by Stam, the author offers an exposition of his novel and elegant 2017 proof as a nice conclusion to this tutorial.

The third contribution, entitled *Landauer's Bound and Maxwell's Demon*, is due to the leading physicist SERGIO CILIBERTO from the Laboratoire de Physique at École Normale Supérieure de Lyon. It describes recent experimental and theoretical progress made in relation to information theory and thermodynamics. The extraordinary wealth of high-quality experimental data at mesoscopic scales implies a complete overhaul of the traditional entropy and irreversibility paradigms in thermodynamics. Together with collaborators, the author performed in 2012 the first experimental test of the *Landauer principle* of 1961, which predicts that a minimum amount of heat $k_B T \ln 2$ is produced when erasing a bit of information, showing for the first time that this limit can be experimentally reached. The author received several scientific awards for his achievements, notably the Prix Jaffé de l'Académie des Sciences de l'Institut de France in 2018 and the EPS Statistical and Non Linear Physics Prize in 2019.

This contribution summarizes in a concise and clear way the basic concepts of *Maxwell's demon* and the *Szilárd engine*, and the resolution in 1982 by C. Bennett of the paradox involved with respect to Clausius' second law. It describes the first experimental implementations of the Maxwell demon, in 2008 with the cooling of atoms in a magnetic trap, in 2010 with the realization of a Szilárd engine with a single microscopic Brownian particle in a fluid in a spiral staircase potential, and in 2015 in electronic circuits. The details of the author's seminal experiment on Landauer's principle, using a colloïdal Brownian particle in a fluid trapped in the double-well potential created by two laser beams, are discussed, as well as the principle's intimate relationship to a generalized form of *Jarzynski's equality*. The survey ends with extensions to the quantum regime, by describing the successful realizations of a quantum Maxwell demon in an NMR set-up in 2016 and in a circuit QED system in 2017, culminating with the recent first experimental verification of the Landauer principle in the quantum case, which used a molecular nano-magnet at a very low temperature.

This volume ends with a perspective on challenges concerning the emerging field of *quantum computation*. It is based on a thoroughly detailed review, entitled *Verification of Quantum Computation: An Overview of Existing Approaches*, by ELHAM KASHEFI, a leading researcher at the Laboratoire d'Informatique de Sorbonne Université and at the School of Informatics at University of Edinburgh, written in collaboration with ALEXANDRU GHIORGHIU, now at the Institute for Theoretical Studies at ETHZ, and with THEODOROS KAPOURNIOTIS at the Department of Physics of the University of Warwick. This work is reprinted from *Theory of Computing Systems* **63**, 715–808 (2019), `https://doi.org/10.1007/s00224-018-9872-3`, under the Creative Commons Attribution 4.0 International License (`http://creativecommons.org/licenses/by/4.0/`).

Over the next five to ten years we will see a state of flux as quantum technologies become part of the mainstream computing landscape. In the meantime we can expect to see quantum computing machines with high variability in terms of architectures and capacities (as we saw when classical computers emerged in the early 1950s). These devices will not be universal in terms of having a simple programming model nor will they be easily applicable to all problems. Adopting and applying such a highly variable and novel technology is both costly and risky for any individual company or research group, as this quantum approach has an acute verification and validation problem: since classical computations cannot scale up to the computational power of quantum mechanics, verifying the correctness of a quantum-mediated computation is challenging; the underlying quantum structure resists classical certification analysis. This contribution covers recently developed techniques to settle these key challenges to make the translation from theory to practice possible. Without this final link the glorious power of quantum technology will not be accessible to us.

This book, by the breadth of topics covered in both the theoretical description and present-day experimental study of the manifold aspects of Information, should be of broad interest to physicists, mathematicians, and philosophers and historians of science. We hope that the continued publication of this series of Proceedings will serve the scientific community, at both the professional and graduate levels.

We thank the COMMISSARIAT À L'ÉNERGIE ATOMIQUE ET AUX ÉNERGIES ALTERNATIVES (Direction de la Recherche Fondamentale), the DANIEL IAGOLNITZER FOUNDATION, the ÉCOLE POLYTECHNIQUE, and the INSTITUT HENRI POINCARÉ for sponsoring the Seminar on which this book is based. Special thanks are due to Chantal DELONGEAS for the preparation of the manuscript.

Saclay & Orsay,
January 2021

BERTRAND DUPLANTIER
Institut de
Physique Théorique
Saclay, CEA, CNRS
Université Paris-Saclay
Gif-sur-Yvette, France
bertrand.duplantier@ipht.fr

VINCENT RIVASSEAU
Laboratoire de
Physique Théorique
CNRS, Univ. Paris-Sud
Université Paris-Saclay
Orsay, France
rivass@th.u-psud.fr

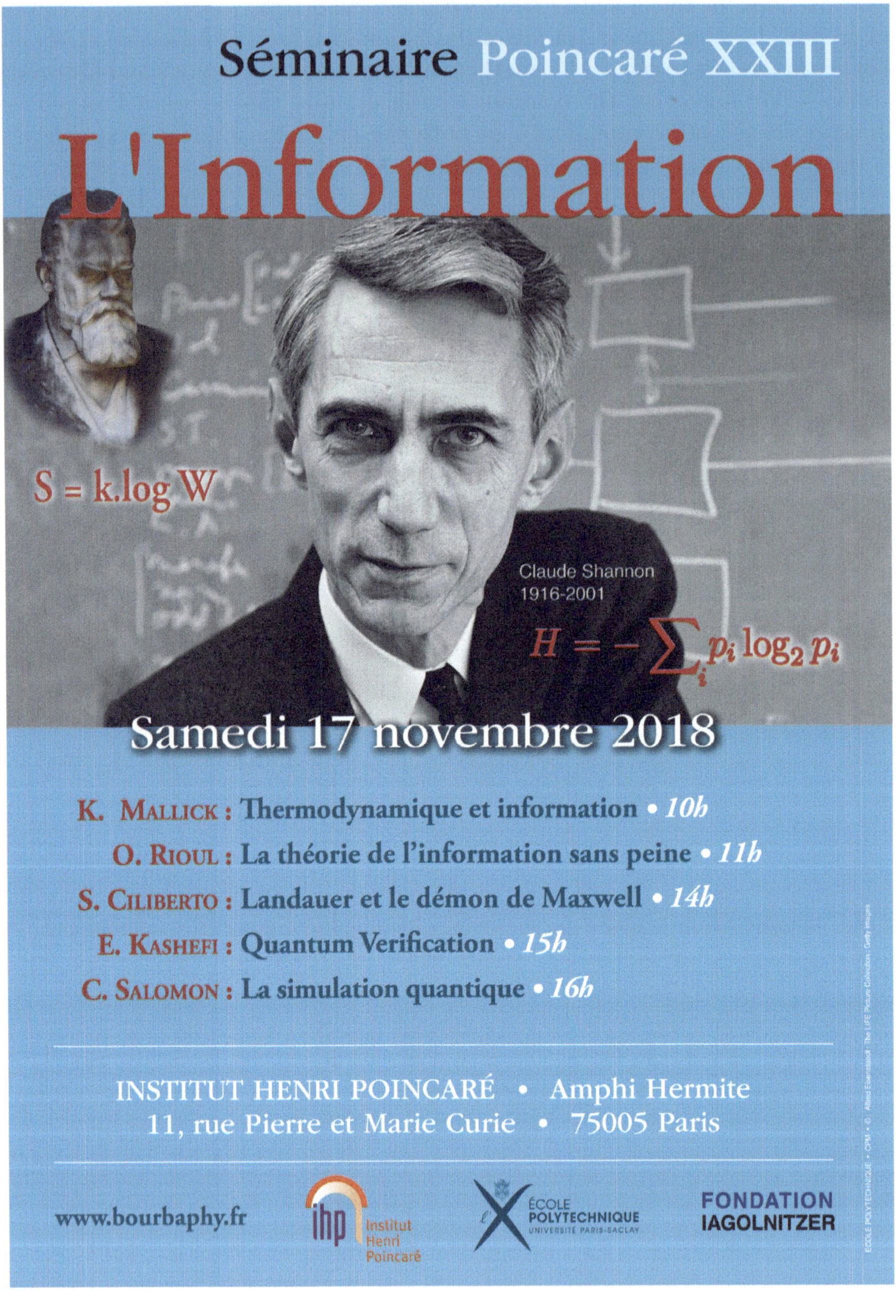
Séminaire Poincaré XXIII
L'Information
$S = k.\log W$
Claude Shannon
1916-2001
$H = -\sum_i p_i \log_2 p_i$
Samedi 17 novembre 2018
K. MALLICK : Thermodynamique et information • 10h
O. RIOUL : La théorie de l'information sans peine • 11h
S. CILIBERTO : Landauer et le démon de Maxwell • 14h
E. KASHEFI : Quantum Verification • 15h
C. SALOMON : La simulation quantique • 16h
INSTITUT HENRI POINCARÉ • Amphi Hermite
11, rue Pierre et Marie Curie • 75005 Paris
www.bourbaphy.fr
ihp Institut Henri Poincaré
ÉCOLE POLYTECHNIQUE
UNIVERSITÉ PARIS-SACLAY
FONDATION IAGOLNITZER

Information Theory, 1–48

Poincaré Seminar 2018

Thermodynamics and Information Theory

Kirone Mallick and Bertrand Duplantier

Abstract. The Laws of classical thermodynamics and the fundamental principles of statistical mechanics are reviewed with an emphasis on their logical structure. We discuss the universality of equilibrium fluctuations (Brownian motion) and describe some contemporary results such as non-equilibrium work identities of Jarzynski and Crooks. A brief review of Shannon entropy reveals the formal analogy between thermodynamics and information theory: these two sciences must be coupled in order to understand and to exorcise Maxwell's demon.

1. Introduction

In his treatise on Thermodynamics [1], Ryogo Kubo mentions a small book by the Czech chemist František Wald (1861–1930), entitled *The Mistress of the World and her Shadow,* a metaphor alluding to energy and entropy. Quoting Robert Emden, Kubo notes that 'in the huge manufactory of natural processes, the principle of entropy occupies the position of manager, for it dictates the manner and method of the whole business, whilst the principle of energy merely does the bookkeeping, balancing credits and debits [2].' Yet, while energy seems to be familiar to all of us, entropy remains a mysterious concept, frequently (mis)used in everyday language as a substitute for chaos, noise, disorder, disorganization or even... business inefficiency [3].

Equilibrium statistical mechanics tells us that entropy relates the microscopic realm to the macroscopic world, by enumerating how many micro-configurations of a system are compatible with our sense-data and with the measurements performed at our scale. It enables us to quantify the loss of information due to coarse-graining from the microscale to the macroscale. To be fair, entropy should be considered as a source of surprise rather than as confusion.

The aim of this article is to recount the subtle ballet of entropy between physics and information theory, choreographed by the puckish demon imagined by Maxwell in 1867, as an incarnation 'of that force that always wills the evil and always produces the good'. Study of entropy will lead us to review the principles of thermodynamics and their underlying statistical basis, with never-ending thermal fluctuations, exemplified by the Brownian motion. By modeling non-equilibrium dynamics, we shall relate entropy to stochastic trajectories. This will lead us to the fluctuation theorem and to the non-equilibrium work identities. The stage will be set to face information theory, to confront the demon and the various attempts to exorcise him (or her).

2. Thermodynamics: A Brief Review

Thermodynamics describes macroscopic properties solid matter, fluid, radiation,... in terms of a small number of macroscopic observables (such as pressure, volume, mass, temperature) when these properties do not vary with time. The laws of thermodynamics ('the two principles') allow us to derive some general relations amongst these properties irrespective of the structure of matter at the atomic scale. Indeed, these principles were established during the XIXth century before the dawn of atomic physics [4, 5, 6, 7, 8, 9, 10].

Thermodynamics can be viewed as the science of *energy conversions*. In order to establish a correct balance, two guiding principles must be respected:

(i) all forms of energy involved must be identified correctly and accounted for;
(ii) different forms of energy are not equivalent. Some energy conversions are free of cost while others come with a fee and need *compensation* (according to Clausius).

Thermodynamics is one of the most elegant branches of physics, but it is also notoriously difficult. This feature has been perfectly emphasized by Onsager (see Fig. 1): 'As in other kinds of bookkeeping, the trickiest questions that arise in the application of thermodynamics deal with the proper identification and classification of the entries; the arithmetics is straightforward' (Onsager, 1967).

2.1. The Two Principles of Thermodynamics

We shall start by reviewing some elementary conversion problems. The simplest example is the conversion of mechanical energy into different forms (kinetic and potential); a ball that falls from a height h reaches the ground with velocity $v^2 = 2gh$ where $g \sim 9.8\, m/s^2$ is the gravity acceleration. This is the content of the celebrated experiments that Galileo is said to have performed from Pisa's leaning tower (see Fig. 2). In this elementary calculation, the friction of air has been neglected: this conversion of potential energy into kinetic energy occurs without a fee (conservation of the total mechanical energy). The free fall of a body can also be used to perform a work $\mathcal{W}$ (e.g., by attaching it to a pulley), the value of which is given by

$$\mathcal{W} = \mathcal{E}_{\text{initial}} - \mathcal{E}_{\text{final}} = -\Delta\mathcal{E} = mgh\,, \tag{1}$$

where $\mathcal{E}$ represents the potential energy.

FIGURE 1. Lars Onsager (1903–1976) obtained the Nobel Prize in Chemistry (1968) for *"the discovery of the reciprocal relations bearing his name, which are fundamental for the thermodynamics of irreversible processes"*. Lars Onsager, 1943. Courtesy of NTNU University Library.

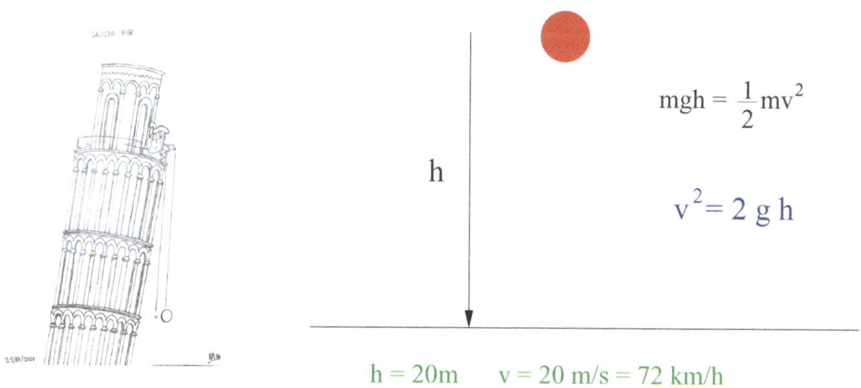

FIGURE 2. The legendary experiment of Galileo in Pisa [Courtesy of Audrey Moché] and its schematic representation.

Using such elementary energy balance arguments one can easily estimate the maximal height that a pole-vaulter can jump on the earth (use the fact that that the highest speed a human being can reach is roughly $10\, m/s$).

The above processes are assumed to be free of dissipation and can be described in purely mechanical terms. In particular, they are perfectly time-reversible: for example, the motion of a pendulum clock allows us to measure time by slicing it into periods of constant duration but it does not tell us what the direction of time is: a movie displaying the oscillations of a dissipation-less pendulum can be run backwards in time without anybody noticing it.

In reality, some dissipation is always present: a ball bouncing on the ground looses energy at each shock and stops after a few bounces. Of course, energy as a whole is conserved, because heat is evolved. This is the content of the first principle of thermodynamics: Heat must be taken into account when doing energy bookkeeping. The work of James Joule established that work and heat are two ways of exchanging energy between a system and its environment. This leads to the *First Law of Thermodynamics*:

$$\boxed{\Delta E = \partial W + \partial Q.} \tag{2}$$

The energy E in this equation is the total internal energy of the system.

In layman's words, the first law states that: **The energy of the universe is constant**.

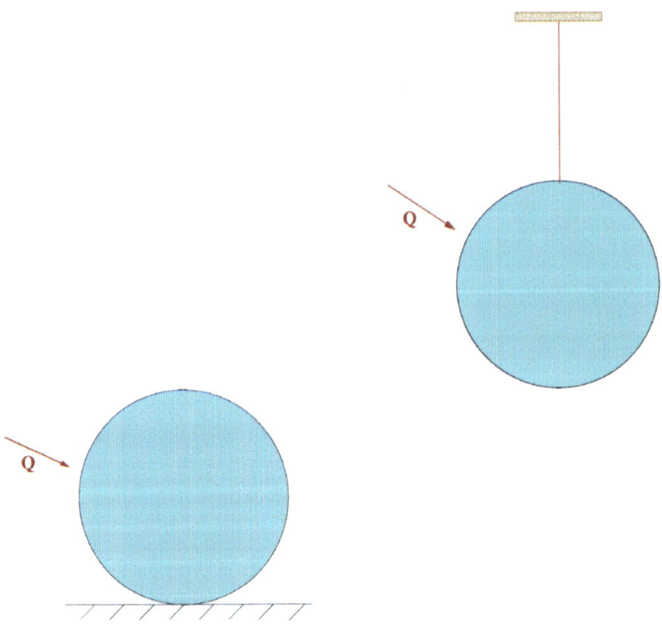

FIGURE 3. The two-sphere puzzle.

Exercise. Energy balance problems can sometimes be rather subtle. Consider two perfectly identical spheres at the same temperature and made of the same material (Fig. 3). One sphere lies on the ground whereas the other is hanging, attached by a rigid thread. The same quantity of heat Q is given to each sphere. Which sphere will be hotter? (We suppose that there is no heat transfer from a sphere to its environment, i.e., ground, air, thread,...) [11].

In presence of dissipation, time-reversibility at macroscopic scale is lost. Projecting the movie of a ball bouncing on the ground backwards in time would display an impossible process: the ball would appear to bounce spontaneously higher and higher by absorbing heat from the ground. Such a process satisfies the first law of thermodynamics but would clearly never happen in reality.

In short, some processes are possible whereas others are not. This can be a very difficult task to detect the hidden flaw in some highly involved mechanisms. How can one *discriminate* between *possible* and *impossible* processes? The solution to this problem is provided by the Second Law of Thermodynamics, elaborated by Carnot in 1824 (see Fig. 4), Clausius (1850) and Kelvin (1851). Two classical formulations of the second law are [4, 5, 7]:

- Clausius formulation: *No process is possible whose* **sole** *result is the transfer of heat from a cooler body to a hotter body.*
- Kelvin-Planck formulation: *No process is possible whose* **sole** *result is the absorption of heat from a reservoir and the conversion of that heat into work.*

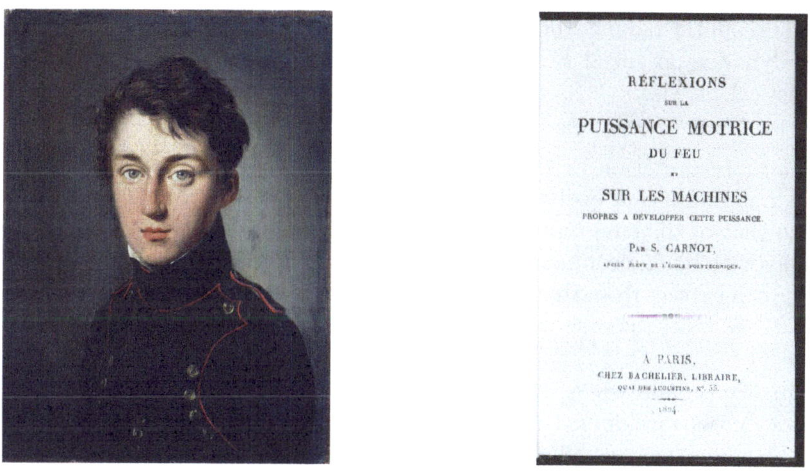

FIGURE 4. Sadi Carnot (1796–1832).

The Clausius and the Kelvin-Planck formulations present two elementary, universal, archetypical forbidden processes. These two statements can be shown to be equivalent and they cover all possible cases; they tell us if a process is possible or not: any impossible process can be proved equivalent to a 'machine' that violates Kelvin-Planck's or Clausius' statement. At this stage, thermodynamics acquires a logical beauty akin to that of classical geometry. Its elegance is perfectly conveyed in the classical books of Fermi [5], Pippard [6] and in the recent textbook of Don Lemons, *Mere Thermodynamics* [7].

The second law was put on a quantitative basis by Clausius, who introduced in 1851 the **entropy**, a state function that measures the degree of irreversibility of a process. This is expressed by Clausius' inequality, which becomes an equality if and only if the process is reversible:

$$\boxed{S_2 - S_1 \geq \int_{1\to 2} \frac{\partial Q}{T}.} \tag{3}$$

The Clausius or Kelvin-Planck statements of the second law can be reformulated in a more formal way: any process that would result in a decrease of the entropy of an isolated system is impossible.

The inequality (3), when applied to the universe considered as a whole, implies that **the entropy of the 'universe' increases**. This sentence can be considered as a popular (albeit with a caveat [12]) statement of the second law.

Although energy is a familiar concept that plays a prominent role in many processes [5, 13, 14], one should never forget entropy that drives secretly many phenomena observed in daily life, for example, the melting of ice. Ultimately, thermodynamic effects are due to the interplay of energy and entropy: A thermal system seeks to minimize its energy while maximizing its entropy at the same time. The subtle balance between these two state functions is encoded in the thermodynamic potential F, called the **free energy** that plays a fundamental role in statistical physics:

$$F = E - TS\,. \tag{4}$$

The interpretation of free energy as maximum available work is classical. Consider a system that evolves from a state A to a state B, both at temperature T equal to that of the environment (see Fig. 5). Suppose that the system exchanges heat only with its environment. Then, because of irreversibility, the *Work*, $\mathcal{W}_{\text{useful}}$, that one can extract from this system is *at most equal to* the decrease of free energy:

$$\boxed{\mathcal{W}_{\text{useful}} \leq F_{\text{initial}} - F_{\text{final}} = -\Delta F.} \tag{5}$$

The equality is valid when the process is reversible. Comparing with equation (1) which is purely mechanical (with no heat transfer) we observe that the role of potential energy is now played by F, and that the equality is replaced by an inequality because of dissipative effects.

Remark. One often considers the work W *we perform* on the system, which is the *opposite* of the work available *from* the system. The inequality (5) then becomes

$$W \geq F_B - F_A = \Delta F. \tag{6}$$

In other words, in order to increase the free energy of an isothermal system by an amount ΔF one has to perform an amount of work at least equal to ΔF. In general, because of irreversibility, the work performed W must be strictly greater than the free-energy variation.

We briefly recall the derivation of the maximal work inequality (6), which requires the two principles of thermodynamics:

- *The first law* states that $\Delta E = W + Q$ (recall that $\mathcal{W}_{\text{useful}}$ is equal to $-W$)
- *The second law:* $\int_{A\to B} \frac{\partial Q}{T} \leq S_B - S_A = \Delta S$ gives that $Q \leq T\Delta S$.

We thus obtain that

$$\Delta F = \Delta E - T\Delta S = W + Q - T\Delta S \leq W.$$

It is useful to define *the dissipated work*, $W_{\text{diss}} = W - \Delta F$, which from the above equation is given by

$$\frac{W_{\text{diss}}}{T} = \Delta S + \left(\frac{-Q}{T}\right) \equiv \Delta S(\text{universe}) \geq 0.$$

The interpretation of this identity is clear: the dissipated work W_{diss}/T *represents the total entropy production, in the universe, by the process.* This entropy production must be non-negative.

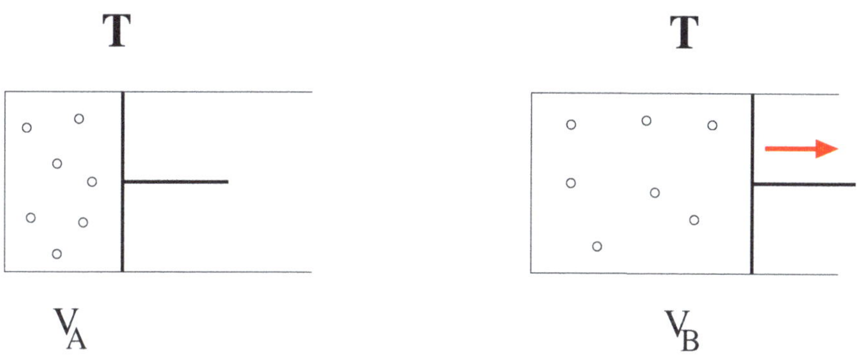

FIGURE 5. An illustration of the maximum work relation on a simple piston-gas system: the piston has been pulled and the volume of the gas has increased from V_A to V_B at constant temperature.

Thermodynamics considers macroscopic observables with well-defined values. However, statistical mechanics predicts that observables are subject to thermal fluctuations around their average, thermodynamic, values. Hence, the work W is also a random variable. Thus, one should be aware that, strictly speaking, thermodynamic identities refer to mean values. In full rigor, the maximal work theorem (5) should be written as

$$\boxed{\langle \mathcal{W}_{\text{useful}} \rangle \leq F_{\text{initial}} - F_{\text{final}} = -\Delta F,} \tag{7}$$

and we emphasize again that the equality can only occur for a reversible process. In a later section of this review, we shall explain that there exists *an exact identity, known as Jarzynski's relation*, which is valid both for reversible and irreversible processes. The classic inequality (7) is a consequence of Jarzynski's relation.

2.2. Molecular Theory of Heat and the Framework of Statistical Mechanics

The works of Maxwell, Boltzmann and Gibbs (Fig. 6) have laid the foundations of statistical mechanics, the microscopic basis of classical thermodynamics.

FIGURE 6. Three fathers of statistical mechanics: James Clerk Maxwell (1831–1879) [Engraved by G.J. Stodart, courtesy of AIP Emilio Segrè Visual Archives], Josiah Willard Gibbs (1839–1903) [Courtesy of AIP Emilio Segrè Visual Archives] and Ludwig Eduard Boltzmann (1844–1906) [Author: Institut of Mathematical Statistics. Source: Archives of the Mathematisches Forschungsinstitut Oberwolfach].

The phase space of a macroscopic system typically made of 10^{23} molecules is huge. Our senses and our measurement devices are sensitive only to a few global and averaged properties. In fact, at the microscopic scale, a system, *even at equilibrium*, evolves continuously from one microstate to another, but most of the time *these different microstates are perceived by us as being the same macroscopic state.* We have indeed access only to extremely *coarse-grained* sense-data, and a tremendous number of microstates are compatible with this data. Thus, coarse-graining from the microscale to the macroscale implies a huge loss of information: entropy quantifies this fact.

More precisely, consider an isolated system of macroscopic volume V with total energy E. We suppose that all microscopic states of the system having energy E are equiprobable: this assumption, known as the 'microcanonical probability ensemble' is a foundation stone of statistical mechanics. (This can be proved rigorously for systems that display sufficiently strong ergodicity.) The thermodynamic entropy S of a microcanonical system enumerates the total number of such microscopic states, $\Omega(E, V)$, and the Boltzmann formula (Fig. 7) states that

$$\boxed{S = k_B \log \Omega} \qquad \text{with } k_B \simeq 1.3810^{-23}. \tag{8}$$

Thus, the determination of Entropy is fundamentally a combinatorial problem.

We shall illustrate this fact by analyzing a simplistic toy model of a perfect gas consisting of N classical distinguishable molecules enclosed in a box of volume

V. A microscopic state is specified by the positions and the velocities of all the molecules. The entropy S of the gas depends on the volume V and the total energy E of the gas. We simplify the discussion further and analyze only the positional configurations (forgetting the velocities – in fact, impulsions ultimately factor out). Let us discretize the box by supposing that the position of an individual molecule inside the box is known up to a precision $\Delta V \ll V$. The total number of micro-configurations is then given by $\Omega = \left(\frac{V}{\Delta V}\right)^N$. Using Boltzmann's formula, we deduce that, at constant energy, the entropy is given by $S = Nk_B \log V$+const. The unwanted constant will drop out if we consider entropy variations. Let us suppose that we double the volume of the box by performing an isothermal expansion $V \to 2V$. Because the temperature is constant, the internal energy of the gas is constant and the variation of entropy, $\Delta S = S_2 - S_1$, is due only to the volume change. We have

$$S_2 - S_1 = Nk_B \log 2. \tag{9}$$

This equation means that each particle contributes to the entropy increase by the amount $k_B \log 2$, which results from the doubling of the available volume. Note that after doubling the volume, a particle can be in two different regions (or 'states'), namely, the left part or the right part of the box with equal probabilities $1/2$.

The combinatorial content of this model is emphasized by making it more abstract and removing all spurious references to gases, particles etc. Consider an assembly of N independent objects, that can occupy m different states that we label $i = 1, 2, \ldots, m$. We assume that state i can be occupied with probability p_i. If $N \gg 1$, we shall have roughly $n_1 = Np_1$ objects in state 1, $n_2 = Np_2$ elements in state 2, etc. The total number or configurations Ω is given by the multinomial coefficient

$$\Omega = \frac{N!}{\prod_{i=1}^m n_i!} = \frac{N!}{\prod_{i=1}^m (Np_i)!}.$$

The corresponding entropy is evaluated by using Stirling's formula and we obtain

$$S = k_B \log \Omega \simeq -Nk_B \sum_{i=1}^m p_i \log p_i. \tag{10}$$

The contribution of each elementary constituent to the total entropy thus can be evaluated as

$$\boxed{S = -k_B \sum_{i=1}^m p_i \log p_i.} \tag{11}$$

Note that the above volume doubling calculation is retrieved by taking $p_1 = p_2 = 1/2$.

This result has far-reaching consequences in fundamental physics. Suppose that the states $i = 1, 2, \ldots, m$ represent m different energy levels $\epsilon_1, \ldots, \epsilon_m$ that can be occupied by N non-interacting objects. Further, consider that the average energy $N\epsilon$ of the system is kept constant, for example, by connecting it to a heat reservoir. As above, we denote by $n_i = p_i N$ the average occupation of state i. We thus have the following two constraints:

$$\sum_{i=1}^{m} p_i = 1, \tag{12}$$

$$\sum_{i=1}^{m} p_i \epsilon_i = \epsilon. \tag{13}$$

The second law of thermodynamics requires that the total entropy S of the ensemble, given by the formula (10), is maximal. As explained in the classical book of Schrödinger [15], we look for the values of the probabilities p_i that maximize S under the constraints (12) and (13). This problem is readily solved (for example, by using Lagrange multipliers) and one finds

$$p_i = \frac{1}{Z} \mathrm{e}^{-\epsilon_i / k_B T}, \tag{14}$$

where the constants Z and T are adjusted to satisfy the constraints. Schrödinger then proves explicitly that T is identical to the physical concept of temperature and that the partition function Z *(Zustandsumme)* is simply related to the thermodynamic free energy (see equation (17) below).

A more standard procedure to study a system thermalized at a given temperature T (i.e., a system in contact with a thermal reservoir at temperature T) is to apply Boltzmann's fundamental equation (8) to the totally isolated entity consisting of the system + reservoir and to eliminate (trace out) the degrees of freedom of the reservoir [13]. The probability of observing a microscopic configuration $\mathcal{C}$ of energy $E(\mathcal{C})$ is then given by the **Boltzmann-Gibbs canonical law:**

$$P_{\mathrm{eq}}(\mathcal{C}) = \frac{\mathrm{e}^{-E(\mathcal{C})/k_B T}}{Z}. \tag{15}$$

The partition function Z which insures that all probabilities sum up to 1 (normalization) is given by

$$Z = \sum_{\mathcal{C}} \mathrm{e}^{-E(\mathcal{C})/k_B T} = \sum_{E} \Omega(E) \mathrm{e}^{-E/k_B T}. \tag{16}$$

The canonical law, which implies a probabilistic description of the microscopic structure of a thermal system is derived from Boltzmann's formula (8) under very general assumptions. The framework of statistical mechanics is laid out by the following relation, deduced from Eqs. (4), (8), (15) and (16) and which links the free energy to the partition function:

$$F = -k_B T \log Z. \tag{17}$$

From this relation, the probabilistic expression of the entropy for a system at temperature T is readily obtained as

$$S = -k_B \sum_{\mathcal{C}} P_{\mathrm{eq}}(\mathcal{C}) \log P_{\mathrm{eq}}(\mathcal{C}), \tag{18}$$

which is identical to the combinatorial expression (11).

The presentation of statistical mechanics varies from one author to another. The standard point of view [13] or the probabilistic/combinatorial approach of Schrödinger [15, 16, 17] are equally valid but their underlying logic is different. The important fact is that statistical mechanics provides us with a systematic procedure to analyze systems at thermal equilibrium:

- Find a suitable microscopic Hamiltonian and describe the microstates of the system.
- Calculate Z and deduce the free energy F.
- Derive from F the thermodynamic properties of the system such as its phase diagram.

Of course, applying this well-defined program to a given problem can be incredibly difficult. Nobody knows how to calculate Z for the three-dimensional Ising model...

FIGURE 7. Ludwig Boltzmann (1844–1906). The celebrated formula for the entropy is inscribed on Boltzmann's grave in Vienna.

The entropy of a mono-atomic classical ideal gas has been calculated by H. M. Tetrode and O. Sakur in 1912 [4, 16]:

$$S = k_B N \log \left[\frac{V}{N} \left(\frac{m k_B T}{2\pi\hbar^2} \right) \right] + \frac{5}{2} k_B N, \tag{19}$$

where m is the mass of a gas particle and $\hbar$ is Planck's constant divided by 2π. For one mole of Helium at 300K, the total entropy is about 100J/K. This expression takes into account quantum indistinguishability and phase-space discreteness.

2.3. Brownian Motion: Equilibrium Is Dynamical

Equilibrium is a dynamical concept: a system in thermal equilibrium keeps on evolving from one microstate to another even if it appears to our imperfect senses to remain in the same macrostate. Thermodynamics deals only with averaged values: it can not account for microscopic fluctuations. Though these fluctuations are usually very minute (of relative order of 10^{-11} for a system containing one mole of matter), they can be detected either by using measuring devices, which are becoming finer and finer, or by studying very small systems. Statistical mechanics allows us to calculate the probability distributions of observables (and not only their averages) and describes perfectly the thermal fluctuations.

The paradigm for thermal fluctuations is Brownian motion discovered by Robert Brown who observed with a microscope, the perpetual, restless, giggling of a pollen grain in water (Fig. 8). This phenomenon is the signature, at our scale, of the granular, discontinuous, structure of matter. It is the experimental footprint of the existence of atoms.

FIGURE 8. Robert Brown (1773–1858) and a sketch of Brownian motion. Photo credits: The Natural History Museum / Alamy Stock Photo.

The theory of Brownian motion was elaborated by Albert Einstein in 1905. The Brownian particle (for example, a grain of pollen) is restlessly shaken by random shocks with the molecules of water. Because of these shocks, the pollen grain undergoes an erratic motion and diffuses with time around its original position: although the position of the Brownian particle does not change with time on average (because of isotropy of space), the quadratic average (i.e., the variance) of the position grows linearly with time:

$$\langle X^2(t) \rangle = 2Dt \,. \tag{20}$$

For a spherical particle of radius a, immersed in a liquid of viscosity η at (absolute) temperature T, the *diffusion constant* D is given, according to Einstein, by

$$D = \frac{RT}{6\pi\eta a\mathcal{N}}, \tag{21}$$

where $\mathcal{N} \simeq 6\,10^{23}$ is the Avogadro number and $R \simeq 8.31$ is the perfect gas constant. This extraordinary formula, discovered by Einstein – and independently by William Sutherland [18] – relates observables D, T, η and a, which are all *macroscopic* quantities, to the number $\mathcal{N}$ of *atoms* in a mole of matter. This relation allowed Jean Perrin to *weigh experimentally an atom of hydrogen* (as he himself stated in his book 'The Atoms'); indeed $1/\mathcal{N}$ is roughly equal to the mass of one atom of hydrogen in grams. In his experiments, Perrin used small latex spheres with $a \sim 0.1\mu m$, immersed in water ($\eta = 10^{-3} kgm^{-1}s^{-1}$) at temperature $T = 300K$. The typical value of D is then $10^{-12}m^2/s$, i.e., the Brownian particles diffuse about one micrometer in one second. Although not strictly macroscopic, such a value of D could be determined by using an optical microscope at the beginning of the twentieth century. The theory of Brownian motion and its experimental verification established beyond any doubt the existence of atoms, considered previously to be a mere hypothesis.

Einstein's formula (21) can be interpreted as the simplest manifestation of the fluctuation–dissipation relation: consider that the pollen grain of size a, immersed in water, is subject to a small *drag force* f_{ext} (suppose, for example, that it is being pulled by an external operator). Because of this force, the pollen acquires a velocity v, and is subject to a frictional force $-\gamma v$ because of the viscosity η of the surrounding water. The friction coefficient γ was calculated by Stokes and is given by $\gamma = 6\pi\eta a$ (assuming the pollen to be a perfect sphere). Balancing the drag force with the frictional force leads to the limiting speed:

$$v_\infty = \sigma f_{\text{ext}} \quad \text{with } \sigma = \frac{1}{6\pi\eta a}. \tag{22}$$

The *susceptibility* σ measures the *linear response* to the external drive f_{ext}. Using this concept of susceptibility, Einstein's relation can be rewritten as:

$$\boxed{D = k_B T\sigma,} \tag{23}$$

$k_B = R/\mathcal{N}$ being Boltzmann's constant. In other words, fluctuations at equilibrium, quantified by D, are proportional to the susceptibility σ which quantifies the linear response to a small external perturbation that drives the system out of equilibrium. There are many good books and articles on Brownian motion and linear response. Some useful references are [13, 19, 20, 18, 21, 22].

2.4. Universality of Brownian Motion: Feynman's Ratchet and Pawl

One reason why Brownian motion was so troublesome to XIXth century physicists was (apart from the fact that they could not find a suitable explanation for it) that the pollen grain was undergoing a kind of perpetual motion even while remaining in contact with a single heat source (the water bath). Moreover, one could conceive a Gedanken-Experiment in which this perpetual motion could be coupled to a mechanical rectifier, a wheel that can rotate only in one direction. Thus, when the Brownian particle would move in one direction, say eastwards, the wheel would rotate, whereas it would stay still if the particle moved westwards (see Fig. 10). This is in essence the celebrated ratchet and pawl model discussed by Feynman in Chap. 46 of his Lectures on Physics, Volume 1 [21]. Feynman rediscovered a model initially proposed by Smoluchowski (see Fig. 9). The second law would then be in trouble, because this rectified motion of the wheel could be used to extract work from a single heat source.

FIGURE 9. Marian Smoluchowski (1872–1917) was a pioneer of statistical physics [Photo: M. Smoluchowski in Lwów, Courtesy of the Jagiellonian Digital Library at Jagiellonian University in Kraków]. Richard Phillips Feynman (1918–1988) made many contributions to fundamental physics, including his ubiquitous diagrams. He also claimed that 'Physics isn't the most important thing. Love is.' R.P. Feynman, June 15, 1955, in front of blackboard (Richard Hartt, photographer), courtesy of Caltech Archives, with kind permission of the Feynman Estate.

In order for the pollen grain to cause rotation of the wheel in the Gedanken-Experiment at a perceptible rate, this wheel must be very small. However, all bodies are subject to thermal fluctuations which typically are inversely proportional to their size. This universal character of thermal fluctuations leads to the resolution of the paradox: the one-way wheel is also subject to intrinsic thermal fluctuations which cause it to move in the forbidden direction. A precise calculation, see, e.g., [25], shows that the two effects (the rotation of the wheel by the Brownian particle and the spontaneous motion in the forbidden direction) perfectly compensate each other and no net rotation of the wheel occurs: 'the second law is saved'.

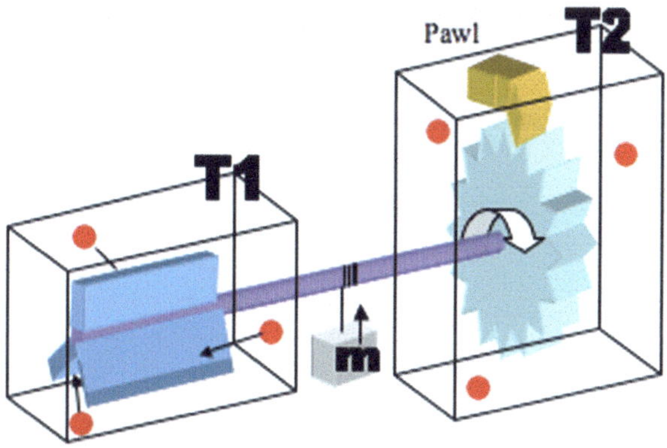

FIGURE 10. A Smoluchowski–Feynman ratchet. Credits: I. Bdkoivis.

2.4.1. Application to molecular motors. The concept of rectification of thermal fluctuations will be useful in non-equilibrium situations and will provide us with a basic model for molecular motors in biological cells.

A significant part of the eucaryotic cellular traffic relies on 'motor' proteins that move in a deterministic way along filaments similar in function to railway tracks or freeways (kinesins and dyneins move along tubulin filaments; myosins move along actin filaments). The filaments are periodic (of period $\sim$ 10nm) and have a fairly rigid structure; they are also polar: a given motor always moves in the same direction. These motors appear in a variety of biological contexts: muscular contraction, cell division, cellular traffic, material transport along the axons of nerve cells, etc.

Molecular motors move by using the ratchet effect: they provide an example of *rectification* of Brownian motion (for reviews see, e.g., [24, 26]). This rectification process relies on an external energy source, provided by ATP (adenosine triphosphate) hydrolysis that enables the motor to undergo transitions between different states, and when these transitions break the detailed balance, a *directed motion* sets in (see Fig. 11). In order to move the motor consumes r ATP fuel molecules per unit time, which are hydrolyzed to ADP + P (adenosine diphosphate + phosphorus):

$$ATP \rightleftharpoons ADP + P\,.$$

The relevant chemical potential is thus given by $\Delta\mu = \mu_{\mathrm{ATP}} - \mu_{\mathrm{ADP}} - \mu_{\mathrm{P}}$.

The principle of the motor is shown in Fig. 12 where the motor is represented by a small particle that can move in a one-dimensional space. At the initial time $t = 0$, the motor is trapped in one of the wells of a periodic asymmetric potential of period a. Between time 0 and t_f, the asymmetric potential is erased and the particle diffuses freely and isotropically. At time t_f, the asymmetric potential is reimposed, the motor slides down in the nearest potential valley and, because of damping, is trapped in one of the wells. The motor has maximal chance to end up

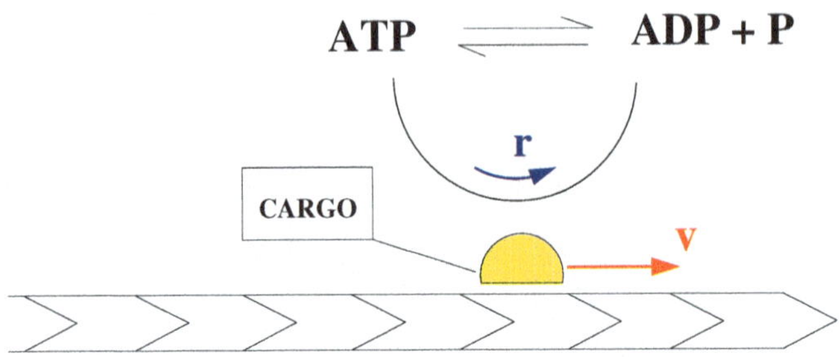

FIGURE 11. Schematic representation of a molecular motor: by hydrolyzing ATP, the motor proceeds along the polar filament and carries a 'cargo' molecule.

in the same well where it was at time $t = 0$. However, it has a small probability to be trapped in the well located to the right and (because of the asymmetry of the potential) an even smaller probability to end up in the left well. In other words, because the potential is asymmetric, the motor has higher chances to slide down towards the right: this leads on average to a net total current.

In general, the motor is subject to an external force f_{ext} which tilts the potential. Besides, when ATP is in excess, the chemical potential $\Delta\mu = \mu_{\text{ATP}} - \mu_{\text{ADP}} - \mu_{\text{P}}$ becomes *positive*. A basic problem is then to determine the velocity of the motor $v(f_{\text{ext}}, \Delta\mu)$ (mechanical current) and the ATP consumption rate $r(f_{\text{ext}}, \Delta\mu)$ (chemical current) as functions of the external mechanical and chemical loads [23, 24, 25, 26, 27].

3. Equilibrium and Non-equilibrium Dynamics

3.1. Markovian Dynamics

An efficient way to describe systems out of equilibrium is to use a probabilistic approach that originates from Einstein's 1905 paper and from Smoluchowski's works at the same period. The idea is to write an evolution equation for the probability $P_t(\mathcal{C})$ for the system to be in the microstate (or configuration) $\mathcal{C}$ at time t. In order to achieve such a description, one has to:

1. Enumerate the microstates $\{\mathcal{C}_1, \mathcal{C}_2, \ldots\}$ of the system. These microstates can form a discrete or a continuous set depending on the problem studied.
2. Specify the transition rates between two configurations. An important and common assumption is that these rates do not depend on the previous history of the system, but only on the configuration $\mathcal{C}$ at time t and on the target configuration $\mathcal{C}'$ at time $t+dt$: this is the *Markovian hypothesis* which amounts

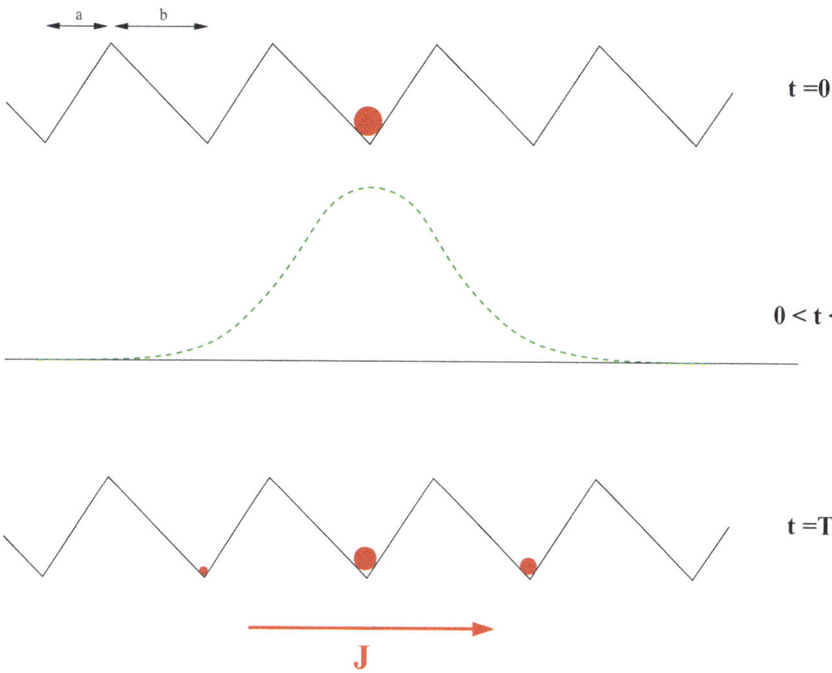

FIGURE 12. The principle of a Brownian ratchet: by inscribing and erasing periodically an asymmetric potential, a directed motion of the particle is induced. In this example, the potential is a saw-tooth function of period $\Lambda = a + b$. Since $b > a$, the slopes are different and the potential is asymmetric. The relative sizes of the probabilities of ending in one of the wells are represented by the sizes of the disks in the lowest figure. The right and left probabilities being different, this leads on average to a net total current J.

to neglecting short time correlations. Thus, one must specify which transitions $\mathcal{C} \to \mathcal{C}'$ are allowed between times t and $t+dt$ with the probability $M(\mathcal{C}', \mathcal{C})dt$.

The evolution of the system can thus be viewed as a path on the configurations network as shown in Fig. 13. The time evolution of $P_t(\mathcal{C})$ is given by a balance equation in this *network* between the incoming flux and the outgoing flux at each configuration:

$$\boxed{\frac{d}{dt} P_t(\mathcal{C}) = \sum_{\mathcal{C}' \neq \mathcal{C}} M(\mathcal{C}, \mathcal{C}') P_t(\mathcal{C}') - \left\{ \sum_{\mathcal{C}' \neq \mathcal{C}} M(\mathcal{C}', \mathcal{C}) \right\} P_t(\mathcal{C}).} \tag{24}$$

This equation is called the **Master equation** in non-equilibrium statistical mechanics. The Master (or Markov) equation (24) is a linear equation in the probability vector P_t, with components given by $\{P_t(\mathcal{C})\}$. It is more convenient to write it in

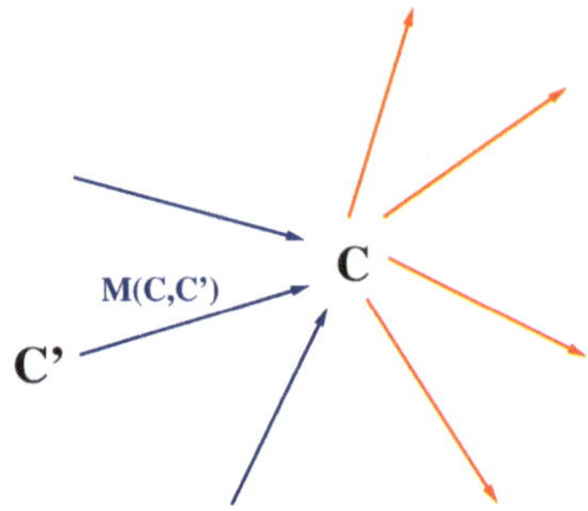

FIGURE 13. A network representation of the transitions in a Markovian dynamics.

a matrix form

$$\frac{dP_t}{dt} = M.P_t\,, \tag{25}$$

where the **Markov operator** M has matrix elements given by $M(\mathcal{C},\mathcal{C}')$ for $\mathcal{C} \neq \mathcal{C}'$ and its diagonal terms are defined as

$$M(\mathcal{C},\mathcal{C}) = -\sum_{\mathcal{C}' \neq \mathcal{C}} M(\mathcal{C}',\mathcal{C})\,. \tag{26}$$

The Markov operator M fully encodes the microscopic dynamics of the system in this probabilistic approach, which already implies a level of coarse-graining. Many physical properties of non-equilibrium systems are obtained by investigating the characteristics of M such as its symmetries or the various types it can assume, etc. Markov operators have the following basic properties:

(i) Non-diagonal terms are non-negative.
(ii) Diagonal terms are negative.
(iii) Equation (26) implies that the sum of the elements of M over any column vanishes. This, in turn, implies the conservation of total probability, i.e., $\sum_{\mathcal{C}} P_t(\mathcal{C})$ is constant.
(iv) A vector P_∞ in kernel of M is a stationary state, i.e.,

$$\frac{dP_\infty}{dt} = 0\,. \tag{27}$$

Under rather general assumptions, it is possible to prove the Perron–Frobenius theorem (see, e.g., [20]) which ensures that: (i) the kernel of M is non-degenerate and therefore the stationary state is unique; (ii) all other eigenvalues of M have strictly negative real parts; the inverse of these real parts correspond to the intrinsic relaxation times of the system towards its stationary state (the imaginary parts characterize the oscillations during relaxation).

Remark. We shall limit our discussion to the Markovian case. The analysis of non-Markovian systems is an active field which requires specific techniques. It often happens, however, that a non-Markovian problem can be embedded into a larger Markovian system by taking into account supplementary degrees of freedom.

The archetypical model of Markovian dynamics is the random walk on a discrete lattice, which is used to explain diffusive behaviour (a classic reference is [19]). We consider a random walker on a one-dimensional discrete lattice with spacing a where he can jump from a site to its two neighbours with probability Γdt between time t and $t + dt$. The Master equation is given by (the transition rate Γ per unit time is normalized to 1):

$$\frac{dP_t(na)}{dt} = P_t((n+1)a) + P_t((n-1)a) - 2P_t(na)\,, \tag{28}$$

and the corresponding Markov operator is nothing but the discrete Laplacian:

$$M = \begin{pmatrix} \ddots & \ddots & \ddots & \ddots & & & \\ 0 & 1 & -2 & 1 & 0 & & \\ 0 & 0 & 1 & -2 & 1 & 0 & \\ & 0 & 0 & 1 & -2 & 1 & \ddots \\ & & \ddots & \ddots & \ddots & \ddots & \ddots \end{pmatrix}.$$

3.2. Connection to Thermodynamics

In the probabilistic approach of non-equilibrium dynamics described above, we made no mention of equilibrium statistical mechanics. Up till now, the discussion has been purely mathematical. However, this description has to be related to thermal physics. The *connection* with thermodynamics is obtained by imposing the condition that the stationary state of the Markovian dynamics P_∞ (Eq. (27)) should be given by the Boltzmann–Gibbs canonical formula (15) which describes a system in thermal equilibrium:

$$P_\infty(\mathcal{C}) = P_{\text{eq}}(\mathcal{C}) = \frac{\mathrm{e}^{-E(\mathcal{C})/k_B T}}{Z}\,. \tag{29}$$

This *matching condition* with thermodynamics imposes, using (26,27), a *global constraint* on the rates in the Markov operator:

$$\sum_{\mathcal{C}'\neq\mathcal{C}} M(\mathcal{C},\mathcal{C}')\mathrm{e}^{-E(\mathcal{C}')/k_B T} = \mathrm{e}^{-E(\mathcal{C})/k_B T}\left\{\sum_{\mathcal{C}'\neq\mathcal{C}} M(\mathcal{C}',\mathcal{C})\right\}. \tag{30}$$

This condition implies that the system described by the stochastic dynamics reaches ultimately a state of thermodynamic equilibrium. For systems far from equilibrium with a non-thermodynamic stationary state, this relation does not hold.

3.3. Time-Reversal Invariance and Detailed Balance

The master equation (24) can be rewritten as follows:

$$\frac{d}{dt}P_t(\mathcal{C}) = \sum_{\mathcal{C}'} \{M(\mathcal{C},\mathcal{C}')\,P_t(\mathcal{C}') - M(\mathcal{C}',\mathcal{C})\,P_t(\mathcal{C})\} = \sum_{\mathcal{C}'} J_t(\mathcal{C},\mathcal{C}')\,, \tag{31}$$

where $J_t(\mathcal{C},\mathcal{C}')$ represents the net local current flowing between the configurations $\mathcal{C}'$ and $\mathcal{C}$ at time t;

$$J_t(\mathcal{C},\mathcal{C}') = M(\mathcal{C},\mathcal{C}')\,P_t(\mathcal{C}') - M(\mathcal{C}',\mathcal{C})\,P_t(\mathcal{C})\,. \tag{32}$$

The local current encodes a balance of ingoing and outgoing currents between $\mathcal{C}'$ and $\mathcal{C}$ that is represented graphically in Fig. 14. The local current is anti-symmetric with respect to the configurations:

$$J_t(\mathcal{C},\mathcal{C}') = -J_t(\mathcal{C}',\mathcal{C})\,. \tag{33}$$

Besides, total probability conservation ($\frac{d}{dt}\sum_{\mathcal{C}} P_t(\mathcal{C}) = 0$) leads to

$$\sum_{\mathcal{C},\mathcal{C}'} J_t(\mathcal{C},\mathcal{C}') = 0\,. \tag{34}$$

This relation is valid for any time t.

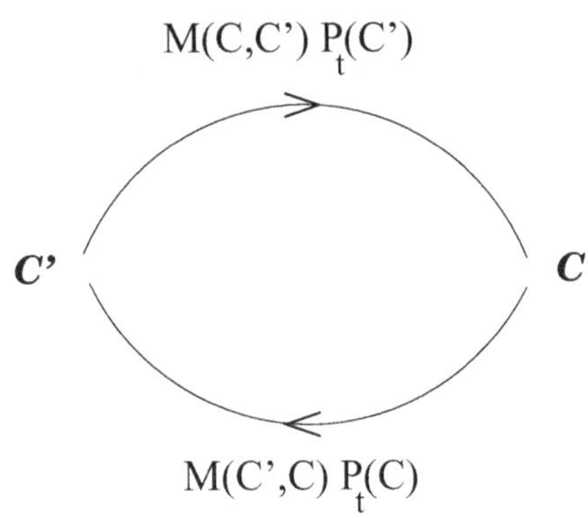

FIGURE 14. The local current between two configurations in a Markovian system.

When $t \to \infty$, the system reaches a steady-state with a stationary probability density. This implies that the total current flowing *to* $\mathcal{C}$ compensates exactly the total current flowing *from* $\mathcal{C}$:

$$\sum_{\mathcal{C}'} J_\infty(\mathcal{C},\mathcal{C}') = 0\,. \tag{35}$$

This global balance condition is nothing but a reformulation of the stationarity condition (27).

Knowing that the microscopic dynamics of the system – represented by an effective Markovian model – is Hamiltonian and that Hamiltonian dynamics is in general time-reversible, Onsager derived the following, much stronger, constraint:

$$M(\mathcal{C},\mathcal{C}')P_{\text{eq}}(\mathcal{C}') = M(\mathcal{C}',\mathcal{C})P_{\text{eq}}(\mathcal{C}). \tag{36}$$

This remarkable relation, known as **detailed balance**, is a consequence of the time-reversal symmetry of the microscopic dynamics of the system. Detailed balance implies that the net local current $J_t(\mathcal{C},\mathcal{C}')$ between two configurations $\mathcal{C}$ and $\mathcal{C}'$ *vanishes at thermodynamic equilibrium.* This very strong condition implies, of course, the vanishing of the global currents (Eq. (35)). The detailed balance condition is a fundamental dynamic property of equilibrium systems that goes beyond the laws of classical thermodynamics. (Magnetic fields and overall rotation effects have been excluded from our discussion for the sake of simplicity. These effects are taken into account in a modified version of detailed balance [20].)

There is an elegant way to express detailed balance as a Hermiticity property of the Markov operator M. Indeed, detailed balance (36) can be rewritten as

$$P_{\text{eq}}(\mathcal{C}')^{1/2}M(\mathcal{C},\mathcal{C}')P_{\text{eq}}(\mathcal{C})^{-1/2} = P_{\text{eq}}(\mathcal{C}')^{-1/2}M(\mathcal{C}',\mathcal{C})P_{\text{eq}}(\mathcal{C})^{1/2}. \tag{37}$$

Introducing the diagonal matrix

$$Q = \begin{pmatrix} \ddots & & \\ & P_{\text{eq}}(\mathcal{C})^{1/2} & \\ & & \ddots \end{pmatrix}, \tag{38}$$

and observing that $M(\mathcal{C},\mathcal{C}') = M^\dagger(\mathcal{C}',\mathcal{C})$, we obtain that detailed balance implies that the operator $\mathcal{W} = Q^{-1}MQ$ is Hermitian:

$$\mathcal{W} = \mathcal{W}^\dagger. \tag{39}$$

A system in thermodynamic equilibrium satisfies detailed balance. Conversely, if the steady state of a system breaks the detailed balance condition, then this steady state is necessarily a non-equilibrium stationary state (NESS). Hence, from a technical point of view, violating detailed balance ensures that the system is out of equilibrium. Moreover, the mathematical description of systems far from equilibrium typically involves non-Hermitian evolution operators.

3.4. Physical Iinterpretation of Detailed Balance

We have said above that detailed balance was derived by Onsager as a consequence of the time-reversibility of the underlying Hamiltonian dynamics. We now prove the reverse assertion: the equilibrium state of a system which satisfies the detailed balance relation (36) is necessarily invariant by time-reversal.

We first recall two properties of any Markovian dynamics which will be useful in the following:

1. The probability of remaining in the state $\mathcal{C}$ during a time interval τ is given by:
$$\lim_{dt\to 0} (1 + M(\mathcal{C},\mathcal{C})dt)^{\frac{\tau}{dt}} = \mathrm{e}^{M(\mathcal{C},\mathcal{C})\tau}. \tag{40}$$
2. The probability of going from $\mathcal{C}$ to $\mathcal{C}'$ during dt is: $M(\mathcal{C}',\mathcal{C})dt$.

Consider a system in its equilibrium state. As we have already emphasized, macroscopic equilibrium is a dynamical state in which the system continuously evolves from one microscopic configuration to another. The frequency with which a given configuration is visited is proportional to its equilibrium probability. We consider a trajectory $\mathcal{C}(t)$ of the system between time $t = 0$ and a final observation time T (see Fig. 15). At $t = 0$ the system is in configuration $\mathcal{C}_0$ with equilibrium probability $P_{\rm eq}(\mathcal{C}_0)$. It jumps from $\mathcal{C}_0$ to $\mathcal{C}_1$ at time $t_1 > 0$ and remains in $\mathcal{C}_1$ till time t_2; at $t_2 > t_1$, it jumps from $\mathcal{C}_1$ to $\mathcal{C}_2$ and remains in $\mathcal{C}_2$ till t_3, etc. More generally, the system jumps from $\mathcal{C}_k$ to $\mathcal{C}_{k+1}$ at time t_{k+1}, for $k = 1, \ldots, n$. The final jump from configuration $\mathcal{C}_{n-1}$ to $\mathcal{C}_n$ occurs at t_n and the system remains in $\mathcal{C}_n$ till the final time T. What is the probability $\Pr\{\mathcal{C}(t)\}$ of observing the trajectory $\mathcal{C}(t)$? Using recursively the two properties recalled above we obtain:
$$\Pr\{\mathcal{C}(t)\} = \mathrm{e}^{M(\mathcal{C}_n,\mathcal{C}_n)(T-t_n)}\,\mathbf{M}(\mathcal{C}_n,\mathcal{C}_{n-1})\,\mathrm{e}^{\mathbf{M}(\mathcal{C}_{n-1},\mathcal{C}_{n-1})(\mathbf{t}_n-\mathbf{t}_{n-1})}\ldots$$
$$\mathrm{e}^{M(\mathcal{C}_2,\mathcal{C}_2)(t_3-t_2)}\,\mathbf{M}(\mathcal{C}_2,\mathcal{C}_1)\,\mathrm{e}^{\mathbf{M}(\mathcal{C}_1,\mathcal{C}_1)(\mathbf{t}_2-\mathbf{t}_1)}\,\mathbf{M}(\mathcal{C}_1,\mathcal{C}_0)\,\mathrm{e}^{\mathbf{M}(\mathcal{C}_0,\mathcal{C}_0)\mathbf{t}_1}\mathbf{P}_{\rm eq}(\mathcal{C}_0). \tag{41}$$

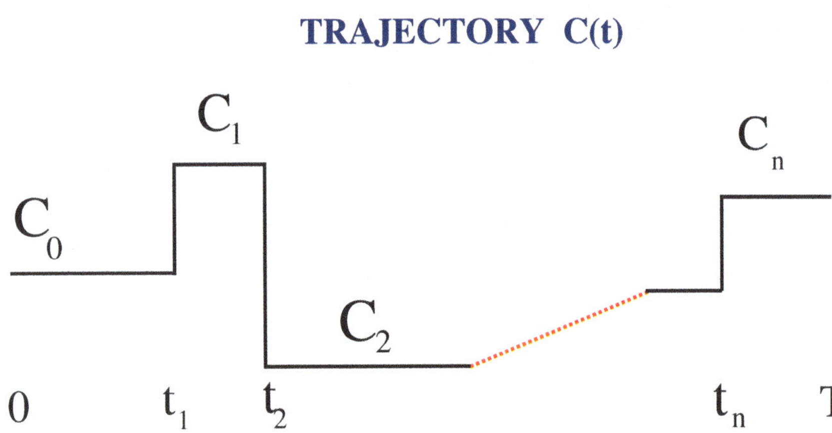

FIGURE 15. A typical trajectory with discrete jumps in a Markovian dynamics.

We now calculate the probability of observing the time-reversed trajectory $\hat{\mathcal{C}}(t) = \mathcal{C}(T-t)$ (see Fig. 16). The system starts at $t = 0$ in configuration $\mathcal{C}_n$ and remains in that configuration till the time $T - t_n$ at which it jumps into $\mathcal{C}_{n-1}$. The next jump from $\mathcal{C}_{n-1}$ to $\mathcal{C}_{n-2}$ occurs at date $T - t_{n-1}$. More generally, the system jumps from $\mathcal{C}_k$ to $\mathcal{C}_{k-1}$ at time $T - t_k$, for $k = n, n-1, \ldots, 1$. At date $T - t_1$, the system reaches the configuration $\mathcal{C}_0$ and remains in it till the final time T. The probability of this trajectory is given by:

$$\Pr\{\hat{\mathcal{C}}(t)\} = \mathrm{e}^{M(\mathcal{C}_0,\mathcal{C}_0)t_1}\,\mathbf{M}(\mathcal{C}_0,\mathcal{C}_1)\,\mathrm{e}^{\mathbf{M}(\mathcal{C}_1,\mathcal{C}_1)(\mathbf{t}_2-\mathbf{t}_1)}$$
$$\ldots\mathrm{e}^{M(\mathcal{C}_{n-1},\mathcal{C}_{n-1})(t_n-t_{n-1})}\,\mathbf{M}(\mathcal{C}_{\mathbf{n-1}},\mathcal{C}_{\mathbf{n}})\,\mathrm{e}^{\mathbf{M}(\mathcal{C}_\mathbf{n},\mathcal{C}_\mathbf{n})(\mathbf{T}-\mathbf{t}_\mathbf{n})}\,\mathbf{P}_{\mathrm{eq}}(\mathcal{C}_\mathbf{n}). \tag{42}$$

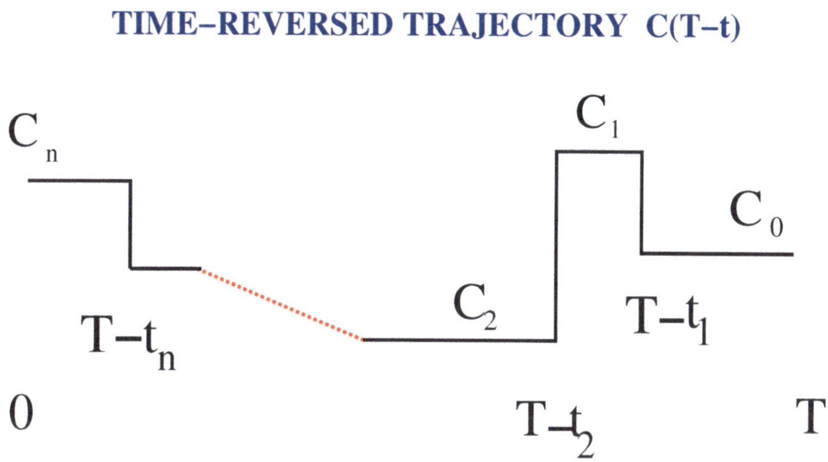

FIGURE 16. The time-reversed trajectory of the trajectory drawn in Fig. 15.

The ratio of the probability of observing a given trajectory (41) to the probability of the time-reversed trajectory (42) is thus given by

$$\frac{\Pr\{\mathcal{C}(t)\}}{\Pr\{\hat{\mathcal{C}}(t)\}} = \frac{M(\mathcal{C}_n,\mathcal{C}_{n-1})M(\mathcal{C}_{n-1},\mathcal{C}_{n-2})\ldots M(\mathcal{C}_1,\mathcal{C}_0)\,P_{\mathrm{eq}}(\mathcal{C}_0)}{M(\mathcal{C}_0,\mathcal{C}_1)\,M(\mathcal{C}_1,\mathcal{C}_2)\,\ldots\quad M(\mathcal{C}_{n-1},\mathcal{C}_n)\,P_{\mathrm{eq}}(\mathcal{C}_n)}. \tag{43}$$

If, in the numerator of this expression, we use recursively the detailed balance condition (36)

$$M(\mathcal{C}_{k+1},\mathcal{C}_k)P_{\mathrm{eq}}(\mathcal{C}_k) = P_{\mathrm{eq}}(\mathcal{C}_{k+1})M(\mathcal{C}_k,\mathcal{C}_{k+1}) \quad \text{for } k = 0,1,\ldots n-1\,,$$

we find that

$$\frac{\Pr\{\mathcal{C}(t)\}}{\Pr\{\hat{\mathcal{C}}(t)\}} = 1\,. \tag{44}$$

We have thus shown that detailed balance implies that the dynamics in the stationary state is time reversible.

3.5. Entropy Production in Markovian Systems

By analogy with equation (18) which gives an expression of the entropy in the canonical ensemble of statistical mechanics, it is possible to define formally a time-dependent 'entropy' function for any Markovian system [20]:

$$S(t) = -\sum_{\mathcal{C}} P_t(\mathcal{C}) \log P_t(\mathcal{C}) \,. \tag{45}$$

Using the Markov equation (31) in terms of the local currents, the time derivative of this function is given by

$$\begin{aligned}\frac{dS(t)}{dt} &= -\sum_{\mathcal{C}} \frac{dP_t(\mathcal{C})}{dt} (\log P_t(\mathcal{C}) + 1) = -\sum_{\mathcal{C},\mathcal{C}'} J_t(\mathcal{C},\mathcal{C}') \log P_t(\mathcal{C}) \\ &= \sum_{\mathcal{C},\mathcal{C}'} J_t(\mathcal{C},\mathcal{C}') \log P_t(\mathcal{C}') \,, \end{aligned} \tag{46}$$

where we have used the global conservation (34). The last equality is obtained by using the antisymmetry of the local currents (33) and by exchanging the role of the dummy variables $\mathcal{C}$ and $\mathcal{C}'$. The expression for the time derivative of $S(t)$ can be written in a more elegant manner by taking the half-sum of the last two equalities:

$$\frac{dS(t)}{dt} = \frac{1}{2} \sum_{\mathcal{C},\mathcal{C}'} J_t(\mathcal{C},\mathcal{C}') \log \frac{P_t(\mathcal{C}')}{P_t(\mathcal{C})} \,. \tag{47}$$

Transforming this expression, we obtain

$$\begin{aligned}\frac{dS(t)}{dt} &= \frac{1}{2} \sum_{\mathcal{C},\mathcal{C}'} J_t(\mathcal{C},\mathcal{C}') \left(\log \frac{P_t(\mathcal{C}')}{P_t(\mathcal{C})} + \log \frac{M(\mathcal{C},\mathcal{C}')}{M(\mathcal{C}',\mathcal{C})} - \log \frac{M(\mathcal{C},\mathcal{C}')}{M(\mathcal{C}',\mathcal{C})} \right) \\ &= \frac{1}{2} \sum_{\mathcal{C},\mathcal{C}'} J_t(\mathcal{C},\mathcal{C}') \log \frac{M(\mathcal{C},\mathcal{C}')P_t(\mathcal{C}')}{M(\mathcal{C}',\mathcal{C})P_t(\mathcal{C})} - \frac{1}{2} \sum_{\mathcal{C},\mathcal{C}'} J_t(\mathcal{C},\mathcal{C}') \log \frac{M(\mathcal{C},\mathcal{C}')}{M(\mathcal{C}',\mathcal{C})} \\ &\equiv \frac{d_i S}{dt} + \frac{d_e S}{dt} \,, \end{aligned} \tag{48}$$

where the first expression in the last equality is called the *entropy production term*, also denoted by σ_i, and the second term the *entropy flux*. The entropy production can be proved, using convexity relations, to be always positive, $\sigma_i \geq 0$, whereas the entropy flux can be positive or negative. At equilibrium, σ_i vanishes identically *because of detailed balance* (Eq. (36)). In the vicinity of equilibrium the linear response theory can be used to show that σ_i decreases towards 0 when the system relaxes to equilibrium. For a system in a non-equilibrium stationary state that does not satisfy detailed balance, σ_i does not vanish but entropy production and entropy flux compensate each other exactly.

4. The Gallavotti–Cohen Fluctuation Theorem for Markovian Thermodynamics

Systems in a non-equilibrium stationary state typically exhibit a non-vanishing macroscopic current J (e.g., a current of particles, or a heat flux,...). Therefore, time-reversal and detailed balance are broken in the mathematical description of the system. The stationary state, which is in the kernel of the Markov operator M, is in general not given by a Boltzmann–Gibbs law (which satisfies detailed balance). In fact, there is no general rule at present which would allow us to calculate the stationary state knowing the external constraints applied to the system. As opposed to the case of thermal equilibrium, there is no general theory of non-equilibrium statistical mechanics. Some general results on systems far from equilibrium have, however, been found that we review now.

4.1. Generalised Detailed Balance

Violation of detailed balance is the source of macroscopic currents which maintain the system far from equilibrium. This violation can be due to different factors:

(i) the existence of an external driving force that pushes the particles in a given direction;
(ii) the presence of reservoirs of unequal chemical potential (or temperature) that generates a current.

The second case is particularly important to model the interaction of a system with its environment and the fluxes that are induced by this interaction. For a system connected to reservoirs there often exists a relation which plays a role similar to that of detailed balance and implies some fundamental properties of the stationary state. This relation is called *generalised detailed balance*. We shall discuss it in the case of a discrete Markovian system [31] which can undergo an elementary transition between two configurations during the interval $(t, t + dt)$. We shall suppose that we are studying an observable Y_t which varies by y at each elementary transition. For each elementary transition, we can specify how Y_t changes:

$$\mathcal{C} \to \mathcal{C}' \text{ and } Y_t \to Y_t + y \quad \text{with probability } M_y(\mathcal{C}', \mathcal{C})dt\,. \tag{49}$$

By *time reversal*, the transition occurs from $\mathcal{C}' \to \mathcal{C}$. Assuming that y is odd (i.e., it changes its sign), we have $Y_t \to Y_t - y$. Finally, we suppose that there exists a constant γ_0 such that transition rates satisfy **the generalised detailed balance condition**:

$$\boxed{M_{+\mathbf{y}}(\mathcal{C}', \mathcal{C}) P_{\text{stat}}(\mathcal{C}) = M_{-\mathbf{y}}(\mathcal{C}, \mathcal{C}')\, e^{\gamma_0 \mathbf{y}}\, P_{\text{stat}}(\mathcal{C}').} \tag{50}$$

For $\gamma_0 = 0$, usual detailed balance is recovered. This relation holds, under general assumptions, for a system in contact with reservoirs that drive it out of equilibrium (in fact, it can be shown to be a consequence of usual detailed balance for the global model obtained by taking into account the system plus the reservoirs).

4.2. Time Reversal and the Gallavotti–Cohen Symmetry

We now investigate the relation between generalised detailed balance and time reversal [31]. We have to modify the calculations done in equations (41), (42), (43) and (44) by taking into account all the factors of the type $e^{\gamma_0 y}$ that appear at each jump in the ratio between the probabilities of forward and time-reversed trajectories.

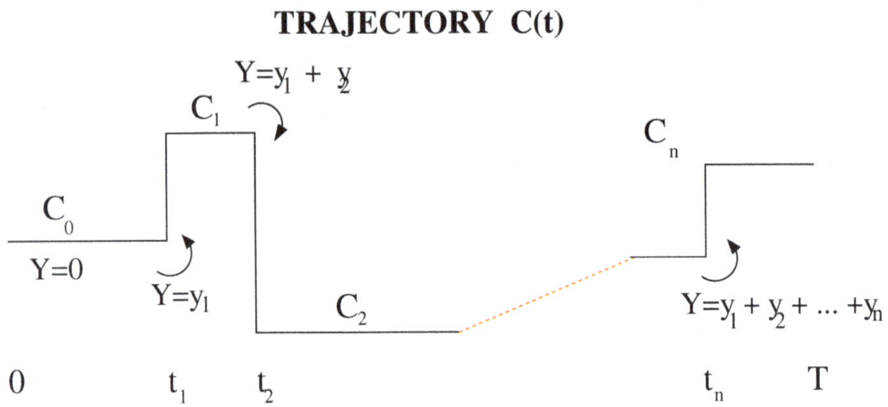

FIGURE 17. A trajectory in a Markovian system; we take into account the variation of the observable Y_t at each jump.

Following the same steps as in Section 3.4, we obtain

$$\frac{\Pr\{\mathcal{C}(t)\}}{\Pr\{\hat{\mathcal{C}}(t)\}} = e^{\gamma_0 Y\{\mathcal{C}(t)\}}, \tag{51}$$

where $Y\{\mathcal{C}(t)\} = y_1 + y_2 + \ldots + y_n$ is the cumulated value when the system follows the trajectory $\mathcal{C}(t)$ between 0 and t (see Fig. 17).

We now recall that Y is odd under time-reversal and therefore we have $Y\{\hat{\mathcal{C}}(t)\} = -Y\{\mathcal{C}(t)\}$. Summing equation (51) over all possible histories between times 0 and t and taking γ to be an arbitrary real number, we obtain

$$\sum_{\mathcal{C}(t)} e^{(\gamma-\gamma_0)Y\{\mathcal{C}(t)\}} \Pr\{\mathcal{C}(t)\} = \sum_{\hat{\mathcal{C}}(t)} e^{-\gamma Y\{\hat{\mathcal{C}}(t)\}} \Pr\{\hat{\mathcal{C}}(t)\}. \tag{52}$$

Because the relation between $\mathcal{C}(t)$ and $\hat{\mathcal{C}}(t)$ is one-to-one, we deduce that

$$\left\langle e^{(\gamma-\gamma_0)Y_t} \right\rangle = \left\langle e^{-\gamma Y_t} \right\rangle, \tag{53}$$

where the sum over all possible paths is interpreted as an average over all possible histories of the process. Using the fact that $\langle e^{\gamma Y_t} \rangle \simeq e^{\mathcal{E}(\gamma)t}$ (see Appendix A), we obtain

$$\mathcal{E}(\gamma - \gamma_0) = \mathcal{E}(-\gamma). \tag{54}$$

Using the Legendre transform of this equation, the **Gallavotti-Cohen fluctuation theorem** [28, 29, 30, 31, 32] for the large deviations function $\Phi(j)$ of the current j is obtained:

$$\boxed{\Phi(j) = \Phi(-j) - \gamma_0 j\,.} \tag{55}$$

Using the definition of the large deviations function (see Appendix A for a brief introduction to large deviations), the fluctuation theorem implies that in the long-time limit

$$\frac{\Pr\left(\frac{Y_t}{t} = j\right)}{\Pr\left(\frac{Y_t}{t} = -j\right)} \simeq \mathrm{e}^{\gamma_0 j}\,. \tag{56}$$

This symmetry property of the large deviations function is valid *far from equilibrium*. This fact has been proved rigorously by various authors in many different contexts (chaotic systems, Markovian dynamics, Langevin dynamics,...).

Remark. In the original works (see, e.g., [31] and references therein), the authors studied the large deviations function for the entropy production σ. The entropy flow is given by (48)

$$\frac{d_e S}{dt} = \frac{1}{2}\sum_{\mathcal{C},\mathcal{C}'} J_t(\mathcal{C},\mathcal{C}') \log \frac{M(\mathcal{C}',\mathcal{C})}{M(\mathcal{C},\mathcal{C}')}\,,$$

where the entropy transfer for each jump is defined as

$$y = \log \frac{M(\mathcal{C}',\mathcal{C})}{M(\mathcal{C},\mathcal{C}')}\,.$$

The increment of the entropy flow at each jump is thus given by y. A property similar to generalised detailed balance is *tautologically true* for y:

$$M_y(\mathcal{C}',\mathcal{C}) = M_{-y}(\mathcal{C},\mathcal{C}')\mathrm{e}^{\gamma_0 y} \quad \text{with} \quad \gamma_0 = 1\,.$$

This relation implies a fluctuation theorem, given by

$$\Phi(\sigma) - \Phi(-\sigma) = -\sigma\,,$$

where $\Phi(\sigma)$ is the large deviations function associated with entropy flow.

5. Non-equilibrium Work Identities

5.1. Jarzynski's Work Theorem

In this section, we describe a remarkable recent result in non-equilibrium statistical physics, which came as a surprise when it was first published by Christopher Jarzynski [33] (Fig. 18). In classical thermodynamics, the work performed on a system in contact with a heat reservoir at temperature T satisfies the relation

$$\langle W \rangle \geq F_B - F_A = \Delta F\,, \tag{57}$$

where F_A is the free energy of the initial state and F_B that of the final state. We point out that the value of the thermodynamic work is in fact an average over many experiments (for example, an operator pulling a piston enclosing a perfect gas from volume V_A to V_B, see Fig. 5). To emphasize this fact we have rewritten here Eq. (6) with the notation $\langle W \rangle$ instead of simply W.

FIGURE 18. Christopher Jarzynski. His first paper on the celebrated Work Identity appeared in Physical Review Letters in 1997 [33]. Courtesy of C. Jarzynski and University of Maryland.

Two decades ago, Christopher Jarzynski found that this classical inequality, well-known since the 19th century, can be deduced from an underlying *remarkable identity* valid for non-equilibrium systems. In the beginning, this identity was proved only for Hamiltonian systems, and was similar in structure, but not equivalent, to predictions derived two decades earlier by German N. Bochkov and Yuriy E. Kuzovlev [34, 35, 36, 37], as later analysed by Jarzynski himself [38] (see also [39]). He (and others) have been extending its validity to more and more cases (such as Markovian dynamics or Langevin systems) [40, 41] and have verified it on exactly solvable models (see, for example, [42, 43, 44] and references therein). Experimental results [45, 46, 47] have also confirmed the Jarzynski relation which is now firmly established and is considered to be one of the few exact results in non-equilibrium statistical mechanics.

Jarzynski's Identity states that

$$\boxed{\left\langle \mathrm{e}^{-\frac{W}{k_B T}} \right\rangle = \mathrm{e}^{-\frac{\Delta F}{k_B T}}.} \tag{58}$$

The average in this equation is taken over a non-equilibrium ensemble of individual trajectories of finite duration t_f.

The precise set up of Jarzynski's identity is as follows. The system has been **prepared in a canonical equilibrium state A at temperature T** and is in this state for $-\infty \le t \le 0$ At time $t = 0$, the state is modified by an external operator according to a *well-defined protocol* $\lambda(t)$ that lasts for a finite period of time $0 \leq t \leq t_f$ (see Fig. 19).

- For $t \leq 0$, $\lambda(0) = \lambda_A$ and the system is *in equilibrium* in the state A.
- Between 0 and t_f, the operator acts on the system by changing a control parameter $\lambda(t)$ according to a fixed well-defined protocol which does not have to be quasi-static and which drives the system far from equilibrium.
- At t_f, the operator **stops** to act and the control parameter is fixed to a value $\lambda(t) = \lambda(t_f) = \lambda_B$ for $t \geq t_f$. We emphasize that the system is *not at equilibrium* at time t_f.

During the whole process, the system remains in contact with a heat-bath at temperature T. After an infinite time, it will reach the thermal equilibrium state B at temperature T. We emphasize that the protocol $\lambda(t)$ is **not assumed** to be 'slow'.

Jarzynski's identity connects data related to a non-equilibrium process (the exponential work average on the left-hand side of the identity) with thermodynamics (the free energy on the right-hand side).

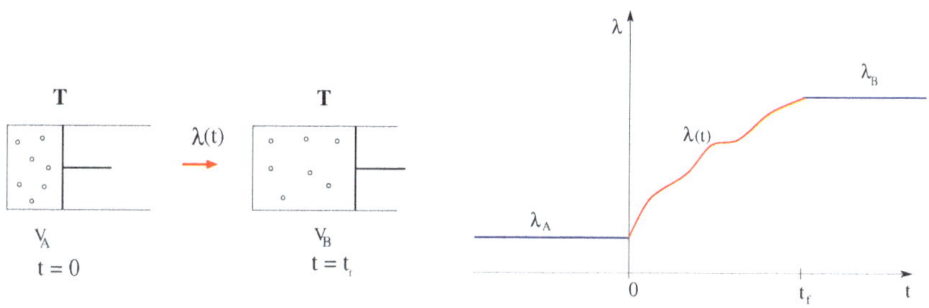

FIGURE 19. Set-up of Jarzynski's formula for the case of a gas in a cylinder. Here $\lambda(t)$ corresponds to the volume: the operator moves the piston according to a well-defined protocol and stops at time t_f when the volume has reached V_B.

Remarks.

1. Using the convexity of the function $x \mapsto \mathrm{e}^{-x}$ and Jensen's inequality, we have
$$\left\langle \mathrm{e}^{-\frac{W}{k_B T}} \right\rangle \geq \mathrm{e}^{-\frac{\langle W \rangle}{k_B T}} .$$
Hence, Jarzynski's work theorem (58) yields the classical inequality (57) for the maximum available work.
2. For Jarzynski's equality to be valid, *there must be* individual trajectories that *do not satisfy* the classical inequality (57), i.e., there must be some realizations for which
$$W < \Delta F , \quad \text{i.e.,} \quad \boxed{W_{\text{useful}} > -\Delta F} .$$

Such special occurrences are called 'transient violations of the second law'. It must be emphasized that the second law is **not** violated because the second law concerns averages and states that the *average* of the performed work is greater than the free energy difference and this remains true. The second law does not say anything about individual behaviour. However, in thermodynamics, we are so used to the fact that individual measurements usually reflect the typical average behaviour that we forget the fact that these two quantities can be different.

5.2. Crooks' Relation

The 'transient violations' of the second law can be quantified thanks to an identity due to Gavin E. Crooks [48, 49, 50] which is more precise than Jarzynski's relation. Let $\lambda^{\mathcal{F}}(t)$ be a protocol of duration t_f that drives the system from V_A to V_B and let $\lambda^{\mathcal{R}}(t) = \lambda^{\mathcal{F}}(t_f - t)$ be the *time reversed* protocol. It is then possible to measure the work done during the forward process and the work done during the reversed process. These quantities are both random variables with probability distributions $\mathrm{P}^{\mathcal{F}}$ and $\mathrm{P}^{\mathcal{R}}$, respectively. The following identity is satisfied by these probability distributions (Crooks, 1999 [48]):

$$\frac{\mathrm{P}^{\mathcal{F}}(W)}{\mathrm{P}^{\mathcal{R}}(-W)} = \mathrm{e}^{\frac{W - \Delta F}{k_B T}} . \tag{59}$$

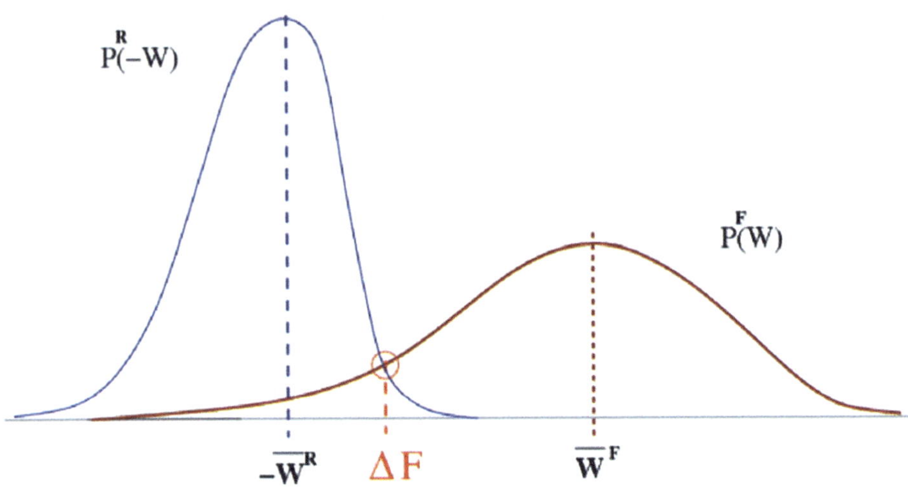

FIGURE 20. Graphical representation of Crooks' relation.

Note the similitude between Crooks' identity and the fluctuation theorem in the form given in Eq. (56) and the proof of Eq. (59) follows similar lines [43]. The main difference with the analysis of Section 4.2 is that the Markov matrix depends on time through the protocol $\lambda(t)$. In order to calculate the ratio of the probabilities for the forward and backward trajectories, a local balance condition is required. We assume, following Crooks [48, 49, 50], that each transition at a given time (and therefore for a given value of λ) satisfies a detailed balance condition:

$$\frac{M_\lambda(\mathcal{C},\mathcal{C}')}{M_\lambda(\mathcal{C}',\mathcal{C})} = \mathrm{e}^{-\frac{E_\lambda(\mathcal{C})-E_\lambda(\mathcal{C}')}{k_B T}},$$

where $E_\lambda(\mathcal{C})$ is the energy of configuration $\mathcal{C}$ for a fixed value of the parameter λ. Calculating, as in Eq. (43), the ratio of the probability of a trajectory to that of the time-reversed trajectory, we obtain, using the local detailed balance condition,

$$\frac{\Pr\{\mathcal{C}(t)\}}{\Pr\{\hat{\mathcal{C}}(t)\}} = \frac{P_{\mathrm{eq},\lambda_0}(\mathcal{C}_0)}{P_{\mathrm{eq},\lambda_n}(\mathcal{C}_n)} \mathrm{e}^{-\frac{1}{k_B T}\sum_{i=1}^n (E_{\lambda_i}(\mathcal{C}_i)-E_{\lambda_i}(\mathcal{C}_{i-1}))}, \tag{60}$$

where the total heat transfered is defined as [48]

$$Q = \sum_{i=1}^n [E_{\lambda_i}(\mathcal{C}_i) - E_{\lambda_i}(\mathcal{C}_{i-1})]. \tag{61}$$

Using this definition, we rewrite Eq. (60) as

$$\frac{\Pr\{\mathcal{C}(t)\}}{\Pr\{\hat{\mathcal{C}}(t)\}} = \mathrm{e}^{\log[P_{\mathrm{eq},\lambda_0}(\mathcal{C}_0)]-\log[P_{\mathrm{eq},\lambda_n}(\mathcal{C}_n)]-\frac{Q}{k_B T}}. \tag{62}$$

Because the variation of energy between the final and the initial configurations is given by

$$\Delta E = E_{\lambda_n}(\mathcal{C}_n) - E_{\lambda_0}(\mathcal{C}_0),$$

we deduce from the first law $\Delta E = Q + W$ that the work performed during the process can be written as

$$\begin{aligned} W &= E_{\lambda_n}(\mathcal{C}_n) - E_{\lambda_0}(\mathcal{C}_0) - \sum_{i=1}^n [E_{\lambda_i}(\mathcal{C}_i) - E_{\lambda_i}(\mathcal{C}_{i-1})] \\ &= \sum_{i=1}^n [E_{\lambda_i}(\mathcal{C}_i) - E_{\lambda_{i-1}}(\mathcal{C}_{i-1})] - \sum_{i=1}^n [E_{\lambda_i}(\mathcal{C}_i) - E_{\lambda_i}(\mathcal{C}_{i-1})] \\ &= \sum_{i=1}^n [E_{\lambda_i}(\mathcal{C}_{i-1}) - E_{\lambda_i}(\mathcal{C}_{i-1})]. \end{aligned} \tag{63}$$

Equations (61) and (63) can be seens as *definitions* of heat and work for a Markov stochastic process with time-dependent transition rates. Besides, we have

$$\frac{P_{\mathrm{eq},\lambda_0}(\mathcal{C}_0)}{P_{\mathrm{eq},\lambda_n}(\mathcal{C}_n)} = \frac{\mathrm{e}^{\beta E_{\lambda_0}(\mathcal{C}_0)}}{Z_{\lambda_0}} \frac{Z_{\lambda_n}}{\mathrm{e}^{\beta E_{\lambda_n}(\mathcal{C}_n)}} = \mathrm{e}^{\frac{\Delta E}{k_B T} - \frac{\Delta F}{k_B T}},$$

where we used the relation $Z = \mathrm{e}^{-\beta F}$ between the partition function and the free energy. Using the first law $\Delta E = Q + W$ again, we conclude that Eq. (60) reduces to

$$\boxed{\frac{\Pr\{\mathcal{C}(t)\}}{\Pr\{\hat{\mathcal{C}}(t)\}} = \mathrm{e}^{\frac{W-\Delta F}{k_B T}},} \tag{64}$$

where W represents the work done along the forward trajectory $\mathcal{C}(t)$. Summing over all trajectories that correspond to the same amount of work and using the fact that W is odd under time reversal, we obtain Crook's relation (59).

We now discuss some consequences of equation (59).

- We note that Jarzynski's relation is a direct consequence of Crooks' equation:

$$\left\langle e^{-\frac{W}{k_B T}} \right\rangle = \int e^{-\frac{W}{k_B T}} \mathrm{P}^{\mathcal{F}}(W)\, \mathrm{d}W = e^{\frac{-\Delta F}{k_B T}} \int \mathrm{P}^{\mathcal{R}}(-W)\, \mathrm{d}W = e^{\frac{-\Delta F}{k_B T}},$$

where for the derivation of the last equality we have used the fact that $\mathrm{P}^{\mathcal{R}}$ is a normalized probability distribution.
- Another consequence of the above analysis that we shall use later, can be obtained as follows. Rewriting Eq. (62) as

$$\sum_{\mathcal{C}(t)} e^{\frac{Q}{k_B T} - \log[P_{\mathrm{eq},\lambda_0}(\mathcal{C}_0)] + \log[P_{\mathrm{eq},\lambda_n}(\mathcal{C}_n)]} \mathrm{Pr}\{\mathcal{C}(t)\} = \sum_{\hat{\mathcal{C}}(t)} \mathrm{Pr}\{\hat{\mathcal{C}}(t)\},$$

sums over all possible paths can be interpreted as averages over all possible histories of the process, and we deduce

$$\langle e^{\frac{Q}{k_B T} - \log[P_{\mathrm{eq},\lambda_0}(\mathcal{C}_0)] + \log[P_{\mathrm{eq},\lambda_n}(\mathcal{C}_n)]} \rangle = 1. \tag{65}$$

Using the convexity of the exponential function, we deduce that (Jensen's inequality)

$$\left\langle \frac{Q}{k_B T} \right\rangle \leq \left\langle \log[P_{\mathrm{eq},\lambda_0}] \right\rangle - \left\langle \log[P_{\mathrm{eq},\lambda_n}] \right\rangle. \tag{66}$$

- We can also calculate the order of magnitude of the probability of a transient violation of the second law, having an amplitude $\zeta > 0$:

$$\begin{aligned} \mathrm{Prob}^{\mathcal{F}}(W < \Delta F - \zeta) &= \int_{-\infty}^{\Delta F - \zeta} \mathrm{P}^{\mathcal{F}}(W)\, \mathrm{d}W = \int_{-\infty}^{\Delta F - \zeta} \mathrm{P}^{\mathcal{R}}(-W)\, e^{\frac{W - \Delta F}{k_B T}}\, \mathrm{d}W \\ &= \int_{-\infty}^{0} \mathrm{P}^{\mathcal{R}}(\zeta - \Delta F - v)\, e^{\frac{v}{k_B T}} e^{-\frac{\zeta}{k_B T}}\, \mathrm{d}v \leq e^{-\frac{\zeta}{k_B T}}. \end{aligned} \tag{67}$$

To derive the last inequality, we have used the fact that $e^{\frac{v}{k_B T}} \leq 1$ for $v \leq 0$ and also that $\int_{-\infty}^{0} \mathrm{P}^{\mathcal{R}} \mathrm{d}v \leq 1$. The probability of a violation of amplitude $\zeta > 0$ is exponentially small with ζ, but again such violations are necessary to ensure the validity of Crooks' and Jarzynski's relations. We observe that for a transient violation to have non-vanishing probability, ζ must be of the order of $k_B T$. On the other hand, $\Delta F \sim N k_B T$, where N is the number of degrees of freedom in the system, which usually are of the order of the Avogadro number. Therefore, transient violations are one part in 10^{23} for macroscopic systems: this is totally unobservable... One has to work with very small systems, such as biophysical objects, to have the chance to observe anything.

Jarzynski's and Crooks' identities allow us to determine equilibrium free energy differences by doing *non-equilibrium experiments.* Many experimental results have been obtained using single-molecule manipulations [45, 46, 47]. As shown in figure 20 representing Crooks' relation (59), the probability distribution $P^F(W)$ of the work W in the forward process and the distribution $P^R(-W)$ of $-W$ in the reversed process cross each other at ΔF. These distributions are determined experimentally in set-ups close to as well as far from equilibrium: Crooks' relation is satisfied in all cases.

6. Information Theory

Thanks to the pioneering work of Claude Elwood Shannon, information theory has become a major discipline of knowledge of the modern age [51]. Many books can introduce the reader to the fundamentals of this subject, from the classic monograph [52] of Shannon and Weaver (Fig. 21) to recent lecture notes of Edward Witten [53], focused on quantum information. A selection of well-known references is [54, 55, 56, 57, 58, 59, 60, 61, 62]. A book that we found enthusiasming is *An introduction to Information Theory. Symbols, Signals and Noise* by John R. Pierce, a colleague and friend of Shannon, who was himself a polymath. Our goal being to focus on the relations between thermodynamics and information, we only recall some basic facts. The reader will find a detailed introduction to information theory in the contribution of Olivier Rioul in this volume.

FIGURE 21. Claude Shannon (1916–2001), father of the information age [Photo credits: Estate of Francis Bello / Science Photo Library]. Right: Shannon's 1948 celebrated article [Courtesy of Manhattan Rare Book Company].

Basic problems of Information Theory can be introduced through elementary examples:

- Suppose I have four objects A, B, C and D. I choose one of them (with the same probability 1/4) and I hide it. You have to determine which one I have selected by asking me *binary questions* (to which I can only answer by yes/no). What is the minimal number $Q_{\min}$ of questions you need to ask?

  Answer: $Q_{\min} = 2$. Devise an optimal strategy and convince yourself that had I started with 2^N objects, you would need to ask $Q_{\min} = N$ questions.
- We now make the problem harder: I still have four objects A, B, C and D, but I select them with unequal probabilities: A with probability 1/2, B with 1/4, C or D with 1/8. What is the minimal number of binary questions that you need to ask me *on average*? Can you devise a strategy? *Answer:* $\langle Q_{\min} \rangle = 7/4$.
- More generally, I have m objects A_i that I select with probability p_i. Is there a lower bound to the average number of binary questions $\langle Q_{\min} \rangle$? Devise an optimal strategy [55].
- Two coins are given: a fair well-balanced coin with a head and a tail, and a fake coin with two heads. I take randomly one of these coins and throw it twice. I record the number of heads obtained. What qualitative information can I deduce about the chosen coin?
- We want to transmit a message using an alphabet made of m letters, $a_1, \ldots,$ a_m, where the letter a_i appears with frequency p_i. Each letter is coded by a binary (0,1) string and we assume that the transmission channel is perfect (noiseless). Can we devise a method to make the code as short as possible [55]?

  Hint: Recall the Morse code, where the most frequent letter E, was represented by a dot •
- Is it possible to transmit information through a noisy channel at a finite rate (we loosely define the rate by the ratio of the length of the message transmitted by the operator with the length of the original message) with an arbitrary small probability of error?

  Answer: YES: this is Shannon's fundamental theorem of information theory (1948). The lower bound of the transmission rate is given by a characteristic property of the channel, called its *capacity* [55, 59].

The solutions to these problems and to many others – that may appear to be totally unrelated to one another – involve a common key concept that Shannon discovered in the late 1940s. Shannon was able to quantify the concept of uncertainty, indeterminacy, (lack of) information, or surprise carried by a probability distribution $p_1, \ldots, p_m$. He proved that uncertainty can be measured in a quantitative manner. Moreover, the measure of information could be expressed by a function $H(\{p_1, \ldots, p_m\})$ that, under very reasonable assumptions, is *unique* (up to an overall normalization constant).

Shannon's H function is given by

$$\boxed{H(\{p_1,\ldots,p_m\}) = -\sum_{i=1}^{m} p_i \log_2 p_i\,,} \tag{68}$$

where $\log_2$ is the base 2 logarithm. H is expressed in bits. In a more compact way, Shannon's H function that measures the lack of information of a probability distribution can be written as

$$H(\{p_i\}) = -\langle \log_2 p \rangle\,. \tag{69}$$

As reported in a famous anecdote, the name of H was suggested to Shannon by von Neumann, in Shannon's words: *My greatest concern was what to call it. I thought of calling it 'information', but the word was overly used, so I decided to call it 'uncertainty'. When I discussed it with John von Neumann, he had a better idea. Von Neumann told me, 'You should call it entropy, for two reasons. In the first place your uncertainty function has been used in statistical mechanics under that name, so it already has a name. In the second place, and more important, no one really knows what entropy really is, so in a debate you will always have the advantage.'*

The Pandora box was open.

7. Thermodynamics and Information: The Maxwell Demon

The mathematical identity between Shannon's H function (68) and the formula for entropy in statistical physics (11) craves for a deep explanation. Edwin T. Jaynes [17] proposed a reformulation of statistical mechanics in which the Shannon entropy (68), re-expressed in suitable units, is taken to be the definition of the thermodynamic entropy. From this point of view, formula (11) is not the result of a combinatorial calculation à la Schrödinger but a starting point. The maximization of the Shannon entropy under appropriate constraints allows one to retrieve the Gibbs ensembles and the potentials of statistical mechanics. Furthermore, Jaynes endeavoured to extend this approach by proposing a minimal entropy production principle to analyze systems far from equilibrium [63]; however, this is only an approximate variational theory that can be considered to be a linearization of more accurate dynamical fluctuation principles [65, 66, 67]. Despite the elegance of Jaynes' work [64], statistical physicists and information theorists continued working in their specific fields, without caring much about the connection between the two entropies.

However, there was one venerable problem where entropy and information had to come face to face, a puzzle that dated from the very beginning of statistical mechanics, Maxwell's demon (see Fig. 22). In his book *Theory of Heat* (1872), Maxwell imagined a thought experiment in which an intelligent being (that Lord Kelvin later christened as a 'demon') is able to violate the second law of thermodynamics by sorting out fast and slow molecules of a gas in a box initially at uniform temperature:

If we conceive of a being whose faculties are so sharpened that he can follow every molecule in its course, such a being, whose attributes are as essentially finite as our own, would be able to do what is impossible to us. For we have seen that molecules in a vessel full of air at uniform temperature are moving with velocities by no means uniform, though the mean velocity of any great number of them, arbitrarily selected, is almost exactly uniform. Now let us suppose that such a vessel is divided into two portions, A and B, by a division in which there is a small hole, and that a being, who can see the individual molecules, opens and closes this hole, so as to allow only the swifter molecules to pass from A to B, and only the slower molecules to pass from B to A. He will thus, without expenditure of work, raise the temperature of B and lower that of A, in contradiction to the second law of thermodynamics.

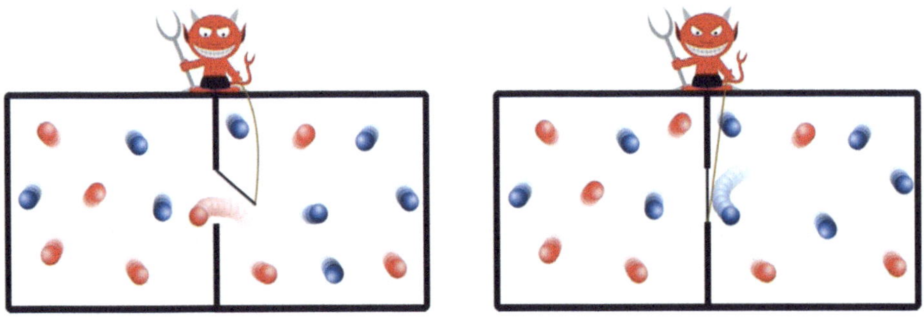

FIGURE 22. Maxwell's demon analyzes the speed of every particle inside the container. By allowing hot particles to enter the left half of the box and keeping the cold particles in the right half, the demon raises the temperature of one chamber in comparison to the other, without expending any work, in contradiction with Clausius' statement of the second law [Courtesy of scienceABS [68]].

For the last 150 years, scientists have struggled to exorcise Maxwell's demon. The various attempts have shed light on the relations between thermodynamics and information [69]. Although the debate is not fully settled yet, significant progress has been made and, recently, Maxwell's demons have appeared in the laboratories, to paraphrase the words of Charles H. Bennett [70], one of the major players in this field. Experimental aspects of the Maxwell's demon are reviewed in [71]. Here, we outline some major theoretical steps that unfold the demon puzzle.

If the demon is an autonomous and inert device, located within the gas, then it is subject to thermal fluctuations and to Brownian motion. This demon could be a trap that opens only in one direction (if it is hit by energetic molecules coming, for example, from the right side of the box). Such a mechanical or electrical rectifier is nothing but a ratchet. As analyzed by Smoluchowski and Feynman (see Section 2.4 and references therein) the universality of Brownian motion would prevent this demon to operate at thermal equilibrium; silly demons are easy to exorcise.

An much more subtle and interesting issue arises if we suppose that the demon can (i) extract information from the gas (via some physical measurement), (ii) record this information, (iii) act accordingly, and (iv) reset itself anew by discarding the acquired information once it has been used. In some sense, the demon has to act in an 'intelligent' way, while remaining analyzable by the laws of physics. In 1929, Leó Szilárd wrote a classic paper, *On the Decrease of Entropy in a Thermodynamic System by the Intervention of Intelligent Beings* [72], in which he proposes a simplified model for an intelligent Maxwell demon, that has become the cornerstone of most of the subsequent studies (see Fig. 23).

A Szilárd engine consists of a single gas particle in a box. The demon determines which half of the box the particle is in (Step A) and inserts a partition (a movable wall) inside the box (Step B). He also attaches a pulley and a mass to the mobile wall. The particle bounces on the wall, pushes it away and performs work by raising the mass (Step C). Because the box is connected to a heat reservoir at temperature T, the expansion of the gas is isothermal. When the wall reaches the left side of the box, it is removed and the cycle is completed (Step D). All told, it seems that the single-particle gas has extracted heat from its environment to perform a total work of $k_B T \log 2$, in apparent contradiction with the second law.

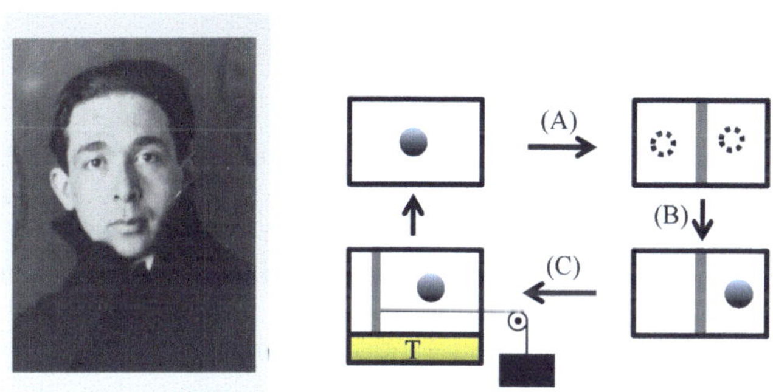

FIGURE 23. Leó Szilárd (1898–1964). Photo credits: [Leó Szilárd Privatdozent ID card photo from Berlin, 1927] Leó Szilárd Papers, MSS32. Special Collections & Archives, UC San Diego Library. On the right: a sketch of Szilárd's engine [From Ref. [73], Courtesy of Jae-Hoon Sim].

Szilárd's engine works by acquiring a binary information (a *bit*) about whether the particle is in the left or right half of the box. Szilárd understood that, to save the second law, the acquisition of this information must cost an entropy production, at least equal to the 'fundamental amount' $k_B \log 2$. However, it was not clear in Szilárd's analysis where this thermodynamic expense was located: in the measurement process, in the recording of the information or in the erasing procedure that closes the cycle?

FIGURE 24. Léon Brillouin (1889–1969) associated thermodynamics with information theory in his attempt to exorcise Maxwell's demon [Courtesy of AIP Emilio Segré Visual Archives, Leon Brillouin collection].

In 1951, Léon Brillouin (Fig. 24), and also, independently, Dennis Gabor, analyzed thoroughly the measurement process by postulating that the demon uses photons to acquire information, that are distributed according to Planck's blackbody law. Brillouin made a bold step: he added, in his entropy balance equations, a contribution coming from the newly discovered Shannon information entropy to the usual thermodynamic entropy. In other words, he argued that information entropy and thermodynamic entropy were directly connected and should be treated on equal footing. The conversion coefficient between information and physical entropy is given by $\mathbf{1\ bit} = \mathbf{k_B \log 2}$, as found by Szilárd. In Brillouin's interpretation, the second law has to be generalized in order to exorcise Maxwell's demon. Thus, after one cycle of the Szilárd engine, the enclosed gas returns to its initial state while the environment has lost a total heat $\Delta Q = -k_b T \log 2$ (corresponding to the work done by the engine) implying a decrease of the total entropy of the universe of $\Delta S = \Delta Q/T = -k_b \log 2$ in contradiction to the second law. However, in Step B (see Fig. 23) an information ΔI has been gained on the position of the molecule. The Shannon entropy corresponding to this information is given by $\Delta I = H = -\frac{1}{2}\log_2\frac{1}{2} - \frac{1}{2}\log_2\frac{1}{2} = 1\text{bit}$. If all sources of entropy are duly recorded on the balance sheet (with the correct conversion factor), we realize that

$$\Delta S + \Delta I \geq 0.$$

The entropy loss of the universe is accompanied by an information gain that compensates it [54, 74, 75].

One important point that was not in Brillouin's analysis was the *physical origin* of the increase of entropy due to acquisition of information. Brillouin and Gabor attributed it to the measurement process which used photons distributed according to the black body radiation. It appeared however that reversible measurement schemes could be devised [79, 80] as shown by Charles H. Bennett (Fig. 25) and that Brillouin's exorcism was not sufficient to get rid of the demon.

In 1961, Rolf Laudauer (Fig. 25) put forward a new principle, relying on the fact that *'information is physical'*. Laudauer introduced the concept of logical irreversibility [76, 77] and analyzed *the process of erasure of information* (a similar idea was explored by Oliver Penrose in [78]). After completing one cycle, the demon has to set back its memory to its original state before starting afresh. According to Landauer, memory erasure is a source of heat and entropy that can never be avoided. The **Landauer principle** *states that the minimum possible amount of energy required to erase one bit of information, known as the Landauer limit is given by* $\mathbf{k_B \log 2}$.

The methods of Sections 4.2 and 5.2 can be used for a microscopic derivation of Landauer's principle [81, 82]. We follow the work of Barbara Piechocinska [81] that gives a microscopic model for information erasure and leads to a bound for the minimal dissipated entropy. A single bit in contact with a thermal reservoir is considered, with two classical states (or levels) denoted by *"zero"* and *"one"*. At $t = 0$, the energy difference, λ_0, between the two levels is zero: the bit can be in state "zero" or "one" with equal probability 1/2. The erasing procedure amounts to increasing the value of λ, and thus performing work on the bit, up to a value $\lambda_n \gg kT$. In the final state, the bit will be in the level "zero" with a probability extremely close to one. This situation belongs to the framework used in Section 5.2 to prove Crooks' relation. Rewriting the inequality (66) with Shannon's H function, we find that the average entropy $\Delta S = -\langle Q\rangle/T$ dissipated by the system in the environment due to the erasure process satisfies:

$$k_B \log 2 \left\{H[P_{\mathrm{eq},\lambda_0}] - H[P_{\mathrm{eq},\lambda_n}]\right\} \leq \Delta S \quad \left(= -\frac{\langle Q\rangle}{T}\right).$$

The initial distribution of the bit is uniform, its Shannon entropy is 1. The final distribution – almost deterministic – is localized in the zero state and its Shannon entropy vanishes. Thus, in order to erase the bit, the entropy of the 'universe' has to increase by at least $k_B \log 2$:

$$\boxed{\Delta S \geq k_B \log 2}.$$

This entropy 'quantum' $k_B \log 2$ is nothing but the conversion factor between Shannon's entropy and Boltzmann's entropy.

Landauer's surmise can now be tested in the laboratory thanks to modern techniques: seminal experiments have been carried out by Sergio Ciliberto, Eric

FIGURE 25. Rolf Landauer (1927–1999) discovered the principle that bears his name, 'information is physical' [Courtesy of AIP Emilio Segré Visual Archives, Physics Today Collection]. Charles Bennett, on the right (born in 1943), is one of the founding fathers of quantum information theory [Courtesy of the Wolf Foundation].

Lutz and coworkers at ENS Lyon [71]; their results and the contributions of other groups are described in this volume.

The story of the demon is not complete yet. Stimulated by the discoveries of Jarzynski and Crooks, the concepts of energy and entropy were applied to stochastic dynamical systems, along the lines sketched in Section 5. This new stochastic thermodynamics [83, 84, 85, 86] provides us with an efficient framework to formulate classical measurement and feed-back processes and to generalize the second law in various settings [87, 88, 89, 90, 91, 92]. We recommend the book of Takahiro Sagawa, based on his PhD work, for a review of the most recent discoveries [93]. A profound and entertaining overview of thermodynamics, information and Maxwell's demon can be found in the comics *Max the Demon vs Entropy of Doom*, by Assa Auerbach and Richard Codor [94].

8. Conclusion

The aim of this presentation is to arouse the curiosity of a non-specialist reader to the various faces of entropy and to provide him or her with fairly up-to-date entries to the literature. Only classical systems have been discussed here. The field of quantum information is huge and, of course, Maxwell's demon has a quantum twin [95] even more impish and 'subtle', but (hopefully) 'not malicious'. Quantum aspects are reviewed in detail in other contributions to this volume: one may also refer to [53] for a 'mini-introduction' and to [96] for a classic reference. Last but not least, let us mention an ultimate haunting spirit, straddling over a box that could swallow our whole universe and our understanding: the solution of Stephen

Hawking's black hole information paradox would require a combination of quantum mechanics and general relativity [97].

Appendix A. Large Deviations and Cumulant Generating Functions

The concept of a large-deviations function is a useful and well-known tool in probability theory. It will be illustrated by the following example.

Let $\epsilon_1, \ldots, \epsilon_N$ be N binary variables where $\epsilon_k = \pm 1$ with probability $1/2$ for $k = 1, \ldots N$. Suppose that the ϵ_k's are independent and identically distributed. Their sum is denoted by $S_N = \sum_1^N \epsilon_k$. We recall:

1. The *law of large numbers* implies that $S_N/N \to 0$ (almost surely).
2. The *central limit theorem* implies that $S_N/\sqrt{N}$ becomes a Gaussian variable of unit variance.

We now quantify the probability that S_N/N takes a non-typical value r, with $-1 < r < 1$. One can show (using the Stirling formula) that in the large N limit,

$$\Pr\left(\frac{S_N}{N} = r\right) \sim \mathrm{e}^{-N\,\Phi(r)}, \tag{70}$$

where $\Phi(r)$ is given by

$$\Phi(r) = \frac{1+r}{2}\log\left(\frac{1+r}{2}\right) + \frac{1-r}{2}\log\left(\frac{1-r}{2}\right) + \log 2\,. \tag{71}$$

The function $\Phi(r)$ is called a *large deviations function.* Because of the law of large numbers, we know that $\Pr(\frac{S_N}{N} = 0)$ tends to 1 when $N \to \infty$. The large deviations function must therefore vanish at $r = 0$, which is indeed the case.

More generally, let Y_t be a random variable (for example, the total charge transported through a system) that depends on time t. We assume that when $t \to \infty$, we have $\frac{Y_t}{t} \to J$, i.e., $\frac{Y_t}{t}$ converges towards its mean value. The random variable $\mathbf{Y_t}$ satisfies a **large deviations principle** if the following identity holds in the large time limit:

$$P\left(\frac{Y_t}{t} = j\right) \sim \mathrm{e}^{-t\Phi(j)},$$

(where the equivalence is to be understood at the level of the logarithms.) The function $\mathbf{\Phi(j)}$ is a **large deviations function** of the rate of production of Y_t. Note that $\Phi(j)$ is positive and vanishes at $j = J$.

Another useful quantity to consider is the moment-generating function of Y_t defined as the average value $\langle \mathrm{e}^{\mu Y_t} \rangle$. Expanding with respect to the parameter μ, we get

$$\log \langle \mathrm{e}^{\mu Y_t} \rangle = \sum_k \frac{\mu^k}{k!} \langle\langle Y^k \rangle\rangle_c\,,$$

where $\langle\langle Y^k\rangle\rangle_c$ is the k-th cumulant of Y_t. In particular, the first two values $\langle\langle Y\rangle\rangle_c$ and $\langle\langle Y^2\rangle\rangle_c$ are equal to the mean and the variance of the random variable Y. In many cases, one can show that in the long time limit, we have

$$\left\langle e^{\mu Y_t}\right\rangle \simeq e^{E(\mu)t} \qquad \text{when} \quad t\to\infty.$$

The function $E(\mu)$ is the cumulant generating function. The previous identity shows that all cumulants of Y_t grow linearly with time and their values are given by the successive derivatives of $E(\mu)$

It can be readily shown that the large deviations function $\Phi(j)$ and the cumulant generating function $E(\mu)$ are related by Legendre transform:

$$\boxed{E(\mu) = \max_j\left(\mu j - \Phi(j)\right).}$$

Indeed, using saddle-point, we obtain for $t\to\infty$,

$$\left\langle e^{\gamma Y_t}\right\rangle = \int \Pr(Y_t) e^{\gamma Y_t} dY_t = t\int \Pr\left(\frac{Y_t}{t} = j\right) e^{\gamma t j} dj \sim \int dj e^{\gamma t j - t\Phi(j)}.$$

Appendix B. Proof of the Jarzynski Formula for Hamiltonian Dynamics

We present here the original proof given by C. Jarzynski (PRL, 1997) [33]. Suppose that, for $t\le 0$, the system is in a state A at thermal equilibrium with its environment at temperature T. Between $0\le t\le t_f$, the coupling to the thermal bath is plugged out. The system is isolated and evolves deterministically according to a **Hamiltonian dynamics**, with Hamiltonian $\mathbf{H}_{\lambda(\mathbf{t})}(\mathbf{p},\mathbf{q})$, which depends on time through the protocol $\lambda(t)$. For $t\ge t_f$, the Hamiltonian remains fixed at $\mathbf{H}_{\lambda_\mathbf{B}}(\mathbf{p},\mathbf{q})$. The system is reconnected to the thermal bath at T and evolves towards the equilibrium thermodynamic state B. We denote by $z=(p,q)$ the phase space coordinate.

The initial distribution on phase space is the canonical distribution with Hamiltonian $\mathbf{H}_{\lambda_\mathbf{A}}$ and $\beta = 1/kT$:

$$\mathrm{P_A}\left(z_0, t=0\right) = \frac{e^{-\beta H_{\lambda_A}(z_0)}}{Z_A}.$$

In the time interval $[0,t_f]$, the system evolves as

$$\dot{p} = -\frac{\partial H_{\lambda(t)}}{\partial q}, \qquad \dot{q} = \frac{\partial H_{\lambda(t)}}{\partial p}, \tag{72}$$

the initial condition $(p,q) = (p_0,q_0)$ being sampled according to P_A.

During this evolution, the **work received** by the system is given by

$$\boxed{W = \int_0^{t_f} \mathrm{d}t\, \frac{\partial H_{\lambda(t)}(z(t))}{\partial t} = \int_0^{t_f} \mathrm{d}t\, \dot{\lambda}(t)\, \frac{\partial H_{\lambda(t)}(z(t))}{\partial \lambda}}.$$

Note that the evolution is deterministic: the only randomness comes from the initial condition. For a Hamiltonian dynamics (72), we have

$$\frac{\partial H_{\lambda(t)}}{\partial t} = \frac{dH_{\lambda(t)}}{dt}$$

and therefore

$$\boxed{W = \int_0^{t_f} \mathrm{d}t\, \frac{\partial H_{\lambda(t)}(z(t))}{\partial t} = H_{\lambda_B}(z(t_f)) - H_{\lambda_A}(z(0))}.$$

For a given value z_0 of $z(0)$, $z_f = z(t_f)$ is uniquely determined. Let us now calculate the exponential average of the work:

$$\begin{aligned}
\langle \mathrm{e}^{-\beta W} \rangle &= \int \mathrm{d}z_0\, \mathrm{P_A}(z_0) \mathrm{e}^{-\beta(H_{\lambda_B}(z_f)) - H_{\lambda_A}(z_0))} \\
&= \int \mathrm{d}z_0\, \frac{\mathrm{e}^{-\beta H_{\lambda_A}(z_0)}}{Z_A} \mathrm{e}^{-\beta(H_{\lambda_B}(z_f)) - H_{\lambda_A}(z_0))} \\
&= \int \mathrm{d}z_0\, \frac{\mathrm{e}^{-\beta H_{\lambda_B}(z_f)}}{Z_A}.
\end{aligned} \tag{73}$$

To conclude, we must make the change of variables $z_0 \to z_f$. This is a one-to-one mapping by the Hamiltonian flow. The key remark is that the Jacobian of this transformation is equal to 1 because of the *Liouville theorem*, we thus have $\mathrm{d}z_0 = \mathrm{d}z_f$.

This concludes the original proof given by Jarzynski:

$$\langle \mathrm{e}^{-\beta \mathbf{W}} \rangle = \frac{1}{Z_A} \int \mathrm{d}z_f \left|\left| \frac{dz_0}{dz_f} \right|\right| \mathrm{e}^{-\beta H_{\lambda_B}(z_f)} = \frac{Z_B}{Z_A} = \mathrm{e}^{-\beta \mathbf{\Delta F}}.$$

This Hamiltonian proof is mathematically rigorous but from the physics point of view, plugging/unplugging the thermal environment seems artificial. In order to overcome this difficulty, one has to give a mechanical representation of heat exchanges. Here are some ways for achieving this task:

- Take the ensemble 'system + heat bath' as a big isolated system, governed by a full Hamiltonian.
- Represent the heat exchanges between the system and the heat bath by a specific (non-Hamiltonian) dynamics: Nosé–Hoover...
- Add to the system's Hamilton equations, a thermal Langevin noise and a dissipative friction term, related by the fluctuation dissipation theorem, that represents the effect of the heat bath.
- Model thermal interaction by a Markovian dynamics.

One should be aware that work theorems are 'meta-theorems': you have to formulate Jarzynski's identity precisely in different contexts and then to give proofs for various settings (more than 15 different proofs are available).

Acknowledgments

We are thankful to Shamlal Mallick for a very careful reading of the manuscript, and to Audrey Moché for drawing Galileo Galilei in Pisa. We thank Erik Wahlström, Helge Holden and the NTNU University Library for providing us with a portrait photo of Lars Onsager. We are also grateful to Bartlomiej Dybiec, Teresa Jaroszewska, Jakub Zakrzewski and Maciej Kluza for their generous help in obtaining Marian Smoluchowski's portraits.

References

[1] R. Kubo, *Thermodynamics.* North Holland, Amsterdam, 1968.

[2] R. Emden, Why do we Have Winter Heating? *Nature* **141**, 908 (1938).

[3] A. Fuchs, *La thermodynamique et l'effet Montesquieu : les usages de l'entropie hors du champ de la thermodynamique.* Publication du Service Enseignements Supérieurs – Didactique de la Chimie (1988).

[4] M.W. Zemansky and R.H. Dittman, *Heat and Thermodynamics.* McGraw Hill, 1981.

[5] E. Fermi, *Thermodynamics.* Dover, New York, 1956.

[6] A.B. Pippard, *The Elements of Classical Thermodynamics.* Cambridge University Press, Cambridge, 1964.

[7] D.S. Lemons, *Mere Thermodynamics.* The John Hopkins University Press, Baltimore, 2009. See also D.S. Lemons, *Thermodynamic Weirdness: From Fahrenheit to Clausius.* MIT Press, Cambridge, MA, 2019.

[8] R.S. Berry, *Three Laws of Nature: A Little Book on Thermodynamics.* Yale University Press, New Haven & London, 2019.

[9] H.B. Callen, *Thermodynamics and an Introduction to Thermostatistics.* John Wiley and Sons, New Jersey, 1985.

[10] A.M. Steane, *Thermodynamics.* Oxford University Press, Oxford, 2017.

[11] If one takes the thermal expansion of the spheres into account, the center of gravity of the sphere on the ground will ascend – thus gaining potential energy – whereas the center of mass of the hanging sphere will descend – thus loosing potential energy. By conservation of total energy, the hanging sphere will be hotter. (I owe this exercise to my colleague Jean-Marc Victor, LPTMC, Paris 6.)

[12] Huw Price, Time's Arrow and Eddington's Challenge. In *Time, Poincaré Seminar 2010*, B. Duplantier (Ed.), Progress in Mathematical Physics, Vol. 63, Birkhäuser, Basel, 2013.

[13] F. Reif, *Statistical Physics: Berkeley Physics Course Vol. 5.* McGraw-Hill, 1967.

[14] A. Greven, G. Keller and G. Warnecke, Eds., *Entropy.* Princeton University Press, Princeton, 2003.

[15] E. Schrödinger, *Statistical Thermodynamics.* Dover, New York, 1989.

[16] J. Machta, Entropy, information, and computation. *Am. J. Phys.* **67**, 1074 (1999).

[17] E.T. Jaynes, Information Theory and Statistical Mechanics. *Phys. Rev.* **106**, 620 (1957); *Phys. Rev.* **108**, 171 (1957). E.T. Jaynes, *Probability Theory: The Logic of Science.* Cambridge University Press, Cambridge, 2003.

[18] B. Duplantier, Brownian Motion, 'Diverse and Undulating'. In: *Einstein, 1905–2005, Poincaré Seminar 2005*, T. Damour, O. Darrigol, B. Duplantier, V. Rivasseau, Eds., Progress in Mathematical Physics, Vol. 47, Birkhäuser, Basel, 2016.

[19] S. Chandrasekhar, Stochastic Problems in Physics and Astronomy. *Rev. Mod. Phys.* **15**, 1 (1943).

[20] N. Van Kampen, *Stochastic Processes in Physics and Chemistry.* North Holland, Amsterdam, 1992.

[21] R.P. Feynman, R.B. Leighton and M. Sands, *Feynman's Lectures in Physics.* Vol. 1, Chapter 46. Basic Books; New Millennium ed., 2011.

[22] H. Spohn, *Large Scale Dynamics of Interacting Particles.* Springer, Berlin, 1991.

[23] M.O. Magnasco, Szilárd's heat engine. *Europhys. Lett.* **33**, 583 (1996).

[24] F. Jülicher, A. Ajdari, J. Prost, Modeling Molecular Motors. *Rev. Mod. Phys.* **69**, 1269 (1997).

[25] C. Jarzynski and O. Mazonka, Feynman's Ratchet and Pawl, an exactly solvable model. *Phys. Rev. E* **59**, 6448 (1999).

[26] R. Dean Astumian and P. Hänggi, Brownian Motors. *Phys. Today* p. 33 (Nov. 2002).

[27] A. Lau, D. Lacoste, K. Mallick, Nonequilibrium Fluctuations and Mechano-chemical Couplings of a Molecular Motor. *Phys. Rev. Lett.* **99**, 158102 (2007); Fluctuation theorem and large deviation function for a solvable model of a molecular motor. *Phys. Rev. E* **78**, 011915 (2008).

[28] G. Gallavotti and E.D.G. Cohen, Dynamical ensembles in stationary states. *J. Stat. Phys.* **80**, 931 (1995).

[29] D.J. Evans and D.J. Searles, The fluctuation theorem, *Adv. Phys.* **51**, 1529 (2002).

[30] J. Kurchan, Fluctuation theorem for stochastic dynamics. *J. Phys. A: Math. Gen.* **31**, 3719 (1998).

[31] J.L. Lebowitz and H. Spohn, A Gallavotti-Cohen Type Symmetry in the Large Deviation Functional for Stochastic Dynamics. *J. Stat. Phys.* **95**, 333 (1999).

[32] G. Gallavotti, Entropy production in nonequilibrium thermodynamics: a point of view. *Chaos* **14**, 680 (2004).

[33] C. Jarzynski, Nonequilibrium equality for free energy differences. *Phys. Rev. Lett.* **78**, 2690 (1997).

[34] G.N. Bochkov and Yu.E. Kuzovlev, General theory of thermal fluctuations in nonlinear systems. *Sov. Phys.–JETP* **45**, 125 (1977) [*Zh. Eksp. Teor. Fiz.* **72**, 238 (1977)].

[35] G.N. Bochkov and Yu.E. Kuzovlev, Fluctuation-dissipation relations for nonequilibrium processes in open systems. *Sov. Phys–JETP* **49**, 543 (1979) [*Zh. Eksp. Teor. Fiz.* **76**, 1071 (1979)].

[36] G.N. Bochkov and Yu.E. Kuzovlev, Nonlinear fluctuation-dissipation relations and stochastic models in nonequilibrium thermodynamics: I. Generalized fluctuation-dissipation theorem. *Physica A* **106**, 443 (1981).

[37] G.N. Bochkov and Yu.E. Kuzovlev, Nonlinear fluctuation-dissipation relations and stochastic models in nonequilibrium thermodynamics: II. Kinetic potential and variational principles for nonlinear irreversible processes. *Physica A* **106**, 480 (1981).

[38] C. Jarzynski, Comparison of far-from-equilibrium work relations. *Comptes Rendus – Physique* **8**, 495 (2007).

[39] G.N. Bochkov and Yu.E. Kuzovlev, Fluctuation-dissipation relations: achievements and misunderstandings. *Phys.-Usp.* **56**, 590 (2013).

[40] C. Jarzynski, Equilibrium free-energy differences from nonequilibrium measurements: A master-equation approach. *Phys. Rev. E* **56**, 5018 (1997).

[41] C. Jarzynski, Rare events and the convergence of exponentially averaged work values. *Phys. Rev. E* **73**, 046105 (2006).

[42] C. Jarzynski, Nonequilibrium work relations: foundations and applications. *Eur. Phys. J. B* **64**, 331 (2008).

[43] E. Boksenbojm, B. Wynants and C. Jarzynski, Nonequilibrium thermodynamics at the microscale: work relations and the second law. *Physica A* **389**, 4406 (2010).

[44] C. Jarzynski, Equalities and Inequalities: Irreversibility and the Second Law of Thermodynamics at the Nanoscale. *Annu. Rev. Cond. Mat. Phys.* **2**, 329 (2011).

[45] J. Liphardt, S. Dumont, S.B. Smith, I. Tinoco and C. Bustamante, Equilibrium information from nonequilibrium measurements in an experimental test of Jarzynski's equality. *Science* **296**, 1832 (2002).

[46] C. Bustamante, J. Liphardt, F. Ritort, The Non-Equilibrium Thermodynamics of Small Systems. *Physics Today*, p. 43 (July 2005).

[47] F. Ritort, Work fluctuations, transient violations of the Second Law and free-energy recovery methods: Perspectives in theory and experiments. In: *Poincaré Seminar 2003, Bose–Einstein Condensation – Entropy*, J. Dalibard, B. Duplantier, V. Rivasseau, Eds., Progress in Mathematical Physics, Vol. 38, Birkhäuser, Basel, 2003. arXiv:cond-mat/0401311.

[48] G.E. Crooks, Nonequilibrium Measurements of Free Energy Differences for Microscopically Reversible Markovian Systems. *J. Stat. Phys.* **90**, 1481 (1998).

[49] G.E. Crooks, Entropy production fluctuation theorem and the nonequilibrium work relation for free energy differences. *Phys. Rev. E* **60**, 2721 (1999).

[50] G.E. Crooks, Path-ensemble averages in systems driven far from equilibrium. *Phys. Rev. E* **61**, 2361 (2000).

[51] J. Gleick, *Information: A History, a Theory, a Flood.* Fourth Estate Ltd, 2012.

[52] C.E. Shannon and W. Weaver, *The Mathematical Theory of Communication.* University of Illinois Press, 1963.

[53] E. Witten, *A Mini-Introduction to Information Theory.* arXiv:1805.11965

[54] L. Brillouin, *La Science et la Théorie de l'information.* Masson, Paris, 1959.

[55] J.R. Pierce, *An Introduction to Information Theory. Symbols, Signals and Noise.* Dover, New York, 1980.

[56] S. Kullback, *Information Theory and Statistics.* Dover, New York, 1997.

[57] R.B. Ash, *Information Theory.* Dover, New York, 1990.

[58] T.M. Cover and J.A. Thomas, *Elements of Information Theory.* John Wiley and Sons, New Jersey, 2006.

[59] O. Rioul, *Théorie de l'information et du codage.* Lavoisier, Paris, 2007.

[60] D.J.C. MacKay, *Information Theory, Inference, and Learning Algorithms.* Cambridge University Press, Cambridge, 2004.

[61] M. Zinsmeister, *Thermodynamic Formalism and Holomorphic Systems.* SMF/AMS Texts and Monographs, Vol. 2, 1996.

[62] M. Mézard and A. Montanari, *Information, Physics and Computation.* Oxford University Press, Oxford, 2009.

[63] E.T. Jaynes, The minimum entropy production principle. *Ann. Rev. Phys. Chem.* **31**, 579 (1980).

[64] W.T. Grandy, Jr., P.W. Milonni, Eds., *Physics and Probability: Essays in Honor of Edwin T. Jaynes.* Cambridge University Press, Cambridge, 1993.

[65] C. Maes and K. Netocny, *Minimum entropy production principle.* Scholarpedia.

[66] L. Bertini, A. De Sole, D. Gabrielli, G. Jona-Lasinio and C. Landim, Macroscopic fluctuation theory. *Rev. Mod. Phys.* **87**, 593 (2015).

[67] B. Derrida, Non-equilibrium steady states: fluctuations and large deviations of the density and of the currents. *J. Stat. Mech.* P07023 (2007).

[68] A. Peshin, What is Maxwell's Demon?. Science ABC, https://www.scienceabc.com/nature/universe/what-is-maxwells-demon.html.

[69] H.S. Leff and A.F. Rex, *Maxwell's Demon 2, Entropy, Classical and Quantum Information, Computing.* IOP Publishing, Bristol, 2003.

[70] C.H. Bennett and B. Schumacher, *Maxwell's Demons Appear in the Lab..* Nikkei Science (Aug 2011).

[71] E. Lutz and S. Ciliberto, Information: From Maxwell's demon to Landauer's eraser. *Phys. Today* **68**, 30 (Sept. 2015) and S. Ciliberto and E. Lutz, *Landauer's Bound and Maxwell's Demon*, in the present volume.

[72] L. Szilárd, On the decrease of Entropy in a Thermodynamic System by the Intervention of Intelligent Beings. *Z. Phys.* **53**, 840 (1929), [Reprinted in [69]].

[73] K.-H. Kim and S. W. Kim. Szilard's information heat engines in the deep quantum regime. *J. Korean Phys. Soc.* **61**, 1187 (2012).

[74] P. Rodd, Some Comments on Entropy and Information. *Am. J. Phys.* **32**, 333 (1964).

[75] K. Maruyama, F. Nori and V. Vedral, Colloquium: The physics of Maxwell's demon and information. *Rev. Mod. Phys.* **81**, 1 (2009).

[76] R. Landauer, Information is Physical. *Phys. Today* **44**, 23, (Mai 1991).

[77] R. Landauer, Irreversibility and Heat Generation in the Computing Process. *IBM J. Res. Dev.* **5**, 183 (1961).

[78] O. Penrose, *Foundations of Statistical Mechanics.* Pergamon Press, Oxford, 1970.

[79] C.H. Bennett, Demons, Engines and the Second Law. *Sci. Am.* **257**, 108 (Nov. 1987).

[80] C.H. Bennett, The thermodynamics of computation, a review. *Int. J. Theor. Phys.* **21**, 905 (1982).

[81] B. Piechocinska, Information erasure. *Phys. Rev. A* **61**, 062314 (2000).

[82] D. Mandal and C. Jarzynski, Work and information processing in a solvable model of Maxwell's demon. *PNAS* **109**, 11641 (2012).

[83] K. Sekimoto, Langevin Equation and Thermodynamics. *Prog. Theor. Phys. Supp.* **130**, 17 (1998).

[84] U. Seifert, Entropy Production along a Stochastic Trajectory and an Integral Fluctuation Theorem. *Phys. Rev. Lett.* **95**, 040602 (2005).

[85] U. Seifert, Stochastic thermodynamics, fluctuation theorems, and molecular machines. *Rep. Prog.* **75**, 126001 (2012).

[86] C. Van den Broeck, *Stochastic thermodynamics: A brief introduction.* Proc. Int. School of Physics Enrico Fermi, Volume 184: Physics of Complex Colloids, C. Bechinger, F. Sciortino and P. Ziherl, Eds. IOS, Amsterdam; SIF, Bologna, 2014.

[87] J.M.R. Parrondo, J.M. Horowitz and T. Sagawa, Thermodynamics of Information. *Nature Phys.* **11**, 131 (2015).

[88] T. Sagawa and M. Ueda, Second Law of Thermodynamics with Discrete Quantum Feedback Control. *Phys. Rev. Lett.* **100**, 080403 (2008).

[89] T. Sagawa and M. Ueda, Minimal Energy Cost for Thermodynamic Information Processing: Measurement and Information Erasure. *Phys. Rev. Lett.* **102**, 250602 (2009); *Phys. Rev. Lett.* **106**, 189901(E) (2011).

[90] T. Sagawa and M. Ueda, Generalized Jarzynski Equality under Nonequilibrium Feedback Control. *Phys. Rev. Lett.* **104**, 090602 (2010).

[91] S. Toyabe, T. Sagawa, M. Ueda, E Muneyuki and M. Sano, Experimental demonstration of information-to-energy conversion and validation of the generalized Jarzynski equality. *Nature Phys.* **6**, 988 (2010).

[92] T. Sagawa and M. Ueda, Fluctuation Theorem with Information Exchange: Role of Correlations in Stochastic Thermodynamics. *Phys. Rev. Lett.* **109**, 180602 (2012).

[93] T. Sagawa, *Thermodynamics of Information Processing in Small Systems.* Springer, Japan, 2013.

[94] A. Auerbach and R. Codor, *Max the Demon vs Entropy of Doom: The Epic Mission of Maxwell's Demon to Face the 2nd Law of Thermodynamics and Save Earth from Environmental Disaster.* Loose Line Productions Inc., 2017.

[95] R.J. Scully, *The Demon and the Quantum.* Wiley-VCH, Singapore, 2010.

[96] M.A. Nielsen and I.L. Chuang, *Quantum Computation and Quantum Information.* Cambridge University Press, Cambridge, 2000.

[97] J. Preskill, Do Black Holes Destroy Information? arXiv:hep-th/9209058; L. Susskind, Three Lectures on Complexity and Black Holes. arXiv:1810.11563.

Kirone Mallick and Bertrand Duplantier
Université Paris-Saclay, CNRS, CEA
Institut de Physique Théorique
CEA/Saclay
91191 Gif-sur-Yvette
France
e-mail: `kirone.mallick@ipht.fr`
`bertrand.duplantier@ipht.fr`

Information Theory, 49–86

Poincaré Seminar 2018

This is IT: A Primer on Shannon's Entropy and Information

Olivier Rioul

I didn't like the term 'information theory'. Claude [Shannon] *didn't like it either. You see, the term 'information theory' suggests that it is a theory about information – but it's not. It's the transmission of information, not information. Lots of people just didn't understand this.* – Robert Fano, 2001

Abstract. What is Shannon's information theory (IT)? Despite its continued impact on our digital society, Claude Shannon's life and work is still unknown to numerous people. In this tutorial, we review many aspects of the concept of entropy and information from a historical and mathematical point of view. The text is structured into small, mostly independent sections, each covering a particular topic. For simplicity we restrict our attention to one-dimensional variables and use logarithm and exponential notations log and exp without specifying the base. We culminate with a simple exposition of a recent proof (2017) of the entropy power inequality (EPI), one of the most fascinating inequalities in the theory.

1. Shannon's Life as a Child

Claude Elwood Shannon was born in 1916 in Michigan, U.S.A., and grew up in the small town of Gaylord. He was a curious, inventive, and playful child, and probably remained that way throughout his life. He built remote-controlled models and set up his own barbed-wire telegraph system to a friend's house [48]. He played horn and clarinet, and was interested in jazz. He was especially passionate about intellectual puzzles, riddles, cryptograms, gadgets and juggling.

He entered the university of Michigan at age 16, where he studied both electrical engineering and mathematics. He would later describe his information theory as "the most mathematical of the engineering sciences" [46].

2. A Noble Prize Laureate

Shannon graduated in 1936. He found an internship position at MIT as an assistant programmer for the "differential analyzer" – an analog machine to solve second-order differential equations – under the supervision of Vannevar Bush, who would become his mentor. Relay switches control the machine, which brings Shannon to a systematic study of the relay circuits. Using his mathematical knowledge, he established the link between circuits and the symbolic formalism of the Boolean algebra. At only 21, his master thesis [40] revolutionized the use of logic circuits by founding digital circuit design theory. It was described as "possibly the most important, and also the most famous, master's thesis of the century"[22].

For his master's work, Shannon received the Alfred Noble prize in 1940. This prize is an award presented by the American Society of Civil Engineers, and has no connection to the better known Nobel prize established by Alfred Nobel. But Shannon's masterpiece was yet to come: *Information theory* – for which he certainly would have deserved the genuine Nobel prize.

3. Intelligence or Information?

Shannon's PhD thesis [41], defended at MIT in 1940, develops an algebra applied to genetics. But without much contact with practitioners of this discipline, his thesis was never published and remained relatively unknown. It must be noted that immediately after receiving his degree he went to work for the Bell telephone laboratories. At this time, Shannon's major concern was what he called "the transmission of intelligence" – what will become later the theory of information. In a letter to Vannevar Bush dated February 16, 1939, he wrote:

> *Off and on I have been working on an analysis of some of the fundamental properties of general systems for the transmission of intelligence, including telephony, radio, television, telegraphy, etc.* [...] *There are several other theorems at the foundation of communication engineering which have not been thoroughly investigated.* [24]

Shannon read the works of Harry Nyquist [34] and Ralph Hartley [25], published in the late 1920s in the *Bell System Technical Journal*, the specialized research journal of the Bell Laboratories. Nyquist had written about the "transmission of intelligence by telegraph" and Hartley's 1928 paper is entitled "transmission of information." Their works will have a decisive influence on Shannon's information theory.

4. Probabilistic, not Semantic

So what is information? Shannon spent ten years (1939–1948), most of it during wartime effort at Bell Laboratories, of intense reflexion about this notion. During this period, he did not publish a single article on the subject – except for a classified memorandum on cryptography in 1945 [42]. He actually used the term 'communication theory', not 'information theory' in most of his work, and first coined the term 'uncertainty' for what would later become Shannon's 'entropy'. The term 'information' or rather 'mutual information' as a mathematical notion in the theory appeared only in the early 1950s in Robert Fano's seminars at MIT [17].

Shannon first deliberately removed the *semantic* questions from the engineering task. A famous paragraph at the very beginning of his seminal 1948 paper reads:

> *"The fundamental problem of communication is that of reproducing at one point either exactly or approximately a message selected at another point. Frequently the messages have meaning; that is they refer to or are correlated according to some system with certain physical or conceptual entities. These semantic aspects of communication are irrelevant to the engineering problem. The significant aspect is that the actual message is one selected from a set of possible messages. The system must be designed to operate for each possible selection, not just the one which will actually be chosen since this is unknown at the time of design."* [43]

Thus, Shannon models the information source as a *probabilistic* device that chooses among possible messages. A message (a sequence of symbols) is a realization of a stochastic process, like a Markov process. In summary, for Shannon, information is probabilistic, not semantic.

Of course, Shannon never said that the semantic aspects are not important. The concept of human intelligence is certainly not purely computational or probabilistic. This perhaps explains why Shannon preferred the term 'communication theory' over 'information theory'.

5. The Celebrated 1948 Paper

Shannon eventually published *A Mathematical Theory of Communication*, in two parts in the July and October issues of Bell System technical journal [43]. As his Bell Labs colleague John Pierce once put it, this paper "came as a bomb – something of a delayed-action bomb"[35]. It is one of the most influential scientific works that was ever published. Few texts have had such an impact in our modern world.

The paper is at the border between engineering and mathematics. At the time, it was not immediately understood by all: On the one hand, most engineers did not have enough mathematical background to understand Shannon's theorems. On the other hand, some mathematicians had trouble grasping the context of communications engineering and found it "suggestive throughout, rather than mathematical," according to the probabilist Joe Doob [13].

The theory is presented in complete form in this single article. It entirely solves the problems of data compression and transmission, providing the fundamental limits of performance. For the first time, it is proved that reliable communications must be essentially digital.

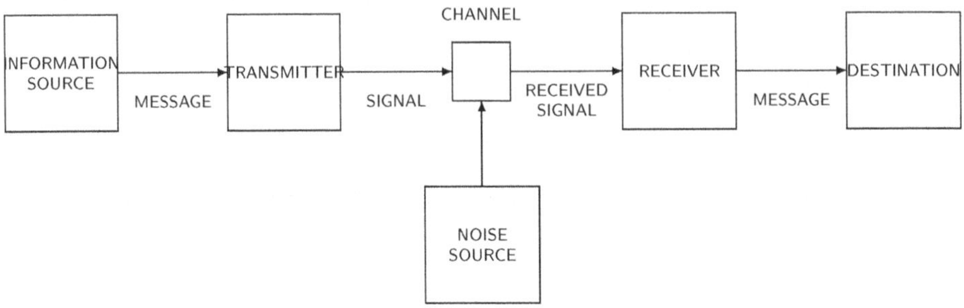

FIGURE 1. Shannon's paradigm, the mother of all models (redrawn from [43]).

Perhaps the most influential part of Shannon's work in all sciences is summarized in the first figure of his 1948 paper: A schematic diagram of a general communication system reproduced in Figure 1, which was called "the mother of all models" in some scientific circles [27]. In this figure, an information source is transmitted over a noisy channel and then received by a recipient. While this scheme seems quite natural today, it was revolutionary: For the first time, we clearly distinguish the roles of source, channel and recipient; transmitter and receiver; signal and noise.

6. Shannon, not Weaver

At the instigation of Shannon's employer Warren Weaver, the 1948 article is republished as a book [44] the following year, preceded by an introductory exposition of Weaver. On this occasion, Shannon's text receives some corrections and some references are updated. But the change that is both the most innocuous and the most important concerns the title: *A mathematical theory of communication* becomes *The mathematical theory of communication.*

Weaver's text, "Recent contributions to the theory of communication," is one of the many contributions to the diffusion of the theory to the general public. A condensed form was published the same year in the popular journal *Scientific American* [57]. Driven by great enthusiasm, Weaver attempts to explain how Shannon's ideas could extend well beyond his initial goals, to all sciences that address communication problems in the broad sense – such as linguistics and social sciences. Weaver's ideas, precisely because they precede Shannon's text in the book, had a tremendous impact: It is likely that many readers came up with the theory while reading Weaver and stopped at Shannon's first mathematical statements.

Even today, the theory is sometimes attributed to Weaver as much as to Shannon, especially in the social sciences. Weaver is often cited as the first author, if not the only author of information theory. It is of course a misinterpretation to attribute the theory to Weaver as well. As Weaver himself declared,

> *"No one could realize more keenly than I do that my own contribution to this book is infinitesimal as compared with Shannon's."* [31]

7. Shannon, not Wiener

Norbert Wiener, the father of cybernetics, has somewhat influenced Shannon. Shannon took Wiener's course in Fourier analysis at MIT [47] and read his wartime classified report "The interpolation, extrapolation and smoothing of stationary time series"[58]. The report is primarily concerned with the linear prediction and filtering problems (the celebrated Wiener filter) but also has some formulation of communication theory as a statistical problem on time series. It was later known to generations of students as the *yellow peril* after its yellow wrappers and the fact that it full of mathematical equations that were difficult to read. Shannon was kind enough to acknowledge that "communication theory is heavily indebted to Wiener for much of its basic philosophy" and that his "elegant solution of the problems of filtering and prediction of stationary ensembles has considerably influenced the writer's thinking in this field."

However, it should be noted that never in Wiener's writings does any precise communication problem appear, and that his use of the term 'information' remained quite loose and not driven by any practical consideration. In his book *Cybernetics* [59], also published in 1948, Wiener deals with the general problems of communication and control. In the course of one paragraph, he considers "the information gained by fixing one or more variables in a problem" and concludes that "the excess of information concerning X when we know Y" is given by a formula identical in form to Shannon's best known formula $\frac{1}{2}\log(1+P/N)$ (see § 33). However, his definition of information is not based on any precise communication problem.

Wiener's prolix triumphalism contrasts with Shannon's discretion. It is likely that the importance of Shannon's formula $\frac{1}{2}\log(1+P/N)$ for which he had made an independent derivation led him to declare:

> *Information theory has been identified in the public mind to denote the theory of information by bits, as developed by C. E. Shannon and myself.* [60]

John Pierce comments:

> *Wiener's head was full of his own work and an independent derivation of* [$\frac{1}{2}\log(1+P/N)$]. *Competent people have told me that Wiener, under the misapprehension that he already knew what Shannon had done, never actually found out.* [35]

8. Shannon's Bandwagon

In the 1950s, Shannon-Weaver's book made an extraordinary publicity. As a result, information theory has quickly become a fashionable field like cybernetics or automation. But, as Shannon himself reckoned, this popularity "carries at the same time an element of danger". While its hard core is essentially a branch of mathematics, the use of exciting words like information, entropy, communication, had led many scientists to apply it indiscriminately to diverse areas such as fundamental physics, biology, linguistics, psychology, economics and other social sciences. So much so that Shannon, in a 1956 editorial entitled "The Bandwagon" [45], warns against the excesses of such popularity:

> [Information theory] *has perhaps been ballooned to an importance beyond its actual accomplishments.* [...] *The subject of information theory has certainly been sold, if not oversold. We should now turn our attention to the business of research and development at the highest scientific plane we can maintain.* [45]

So let us now turn our attention to mathematics.

9. An Axiomatic Approach to Entropy

Entropy is perhaps the most emblematic mathematical concept brought by Shannon's theory. A well-known derivation of Shannon's entropy [43] follows an axiomatic approach where one first enunciates a few desirable properties and then derives the corresponding mathematical formulation. This offers some intuition about a "measure of information". Several variants are possible based on the following argument.

Consider any event with probability p. How should behave the corresponding amount of information $i(p)$ as a function of p? First, the event should bring all the more information as it is unlikely to occur; second, independent events should not interfere, the corresponding amounts of information simply add up. Therefore, two desirables properties are:

(a) $i(p) \geq 0$ is a decreasing function of p;
(b) for any two independent events with probabilities p and q, $i(pq) = i(p) + i(q)$.

Here $i(p)$ can also be interpreted as a measure of "surprise", "unexpectedness", or "uncertainty" depending on whether the event has or has not yet occurred.

Let n be a positive integer and r be the rank of the first significant digit of p^n so that $10^{-r} \geq p^n \geq 10^{-(r+1)}$. Applying (a) and (b) several times we obtain $r \cdot i(1/10) \leq n \cdot i(p) \leq (r+1)\, i(1/10)$, that is,

$$\frac{r}{n} \leq c \cdot i(p) \leq \frac{r}{n} + \frac{1}{n}, \tag{1}$$

where c is constant independent of r and n. Now since the function $\log(1/p)$ satisfies the same properties (a), (b) above, it also satifies

$$\frac{r}{n} \leq c' \cdot \log \frac{1}{p} \leq \frac{r}{n} + \frac{1}{n}, \tag{2}$$

where c' is another constant independent of r and n. It follows from (1) and (2) that

$$\left| c \cdot i(p) - c' \cdot \log \frac{1}{p} \right| \leq \frac{1}{n}. \tag{3}$$

Letting $n \to +\infty$ we obtain that $i(p)$ is proportional to $\log(1/p)$, where the constant of proportionality can be arbitrary. Since the choice of the constant amounts to specifying the base of the logarithm (see § 10), we can simply write

$$i(p) = \log \frac{1}{p}. \tag{4}$$

Now consider the case of a random variable X with probability distribution $p(x)$. The amount of information of an elementary event $X = x$ is then $\log \frac{1}{p(x)}$. Therefore, the *average amount of information* about X is given by the expected value:

$$H(X) = \sum_x p(x) \log \frac{1}{p(x)}. \tag{5}$$

This is Shannon's *entropy* $H(X)$ of the random variable X having distribution $p(x)$.

The notation $H(X)$ may be a little confusing at first: This is not a function of X but rather of its probability distribution $p(x)$. Some authors write $H(p)$ in place of $H(X)$ to stress the dependence on the probability distribution $p(x)$.

One often sees the equivalent formula

$$H(X) = -\sum_x p(x) \log p(x) \tag{6}$$

which is essentially a matter of taste. Note, however, that since probabilities $p(x)$ lie between 0 and 1, the above expression is *minus* the sum of *negative* quantities, whereas (5) is simply the sum of positive quantities.

10. Units of Information

The base of the logarithm in (5) can be chosen freely. Since a change of base amounts to a multiplication by a constant, it specifies a certain *unit* of information.

Suppose, e.g., that X takes M equiprobable values $x = 0, 1, 2, \ldots, M-1$, so that $p(x) = 1/M$ in (5). Then Shannon's entropy is simply the logarithm of the number of possible values:

$$H(X) = \log M. \tag{7}$$

If the values x are expressed in base 10, a randomly chosen m-digit number between 0 and $10^n - 1$ corresponds to $M = 10^n$. With a logarithm to base 10, the entropy is simply the number $m = \log_{10} M$ of decimal digits. Similarly, a randomly chosen m-digit number in base 2 (between 0 and $2^n - 1$) gives an entropy of m binary digits. This generalizes to any base.

With the emergence of computers, the base 2 is by far the most used in today's technology. Accordingly, the entropy is often expressed with a logarithm to base 2. The corresponding unit is the *bit*, a contraction of binary digit. Thus M possible values correspond to $\log_2 M$ bits. It was Shannon's 1948 paper [43] that introduced the word *bit* for the very first time – a word widely used today.

While a bit (a binary digit) is either 0 or 1, the entropy $H(X)$, expressed in bits, can take any positive value. For example, $\log_2 3 = 1.58496\ldots$ bits. Here the word bit (as a unit of information) can be thought of as the contraction of "binary unit" rather than of "binary digit". Similarly, with base 10, the unit of information is the *dit* (decimal unit). For natural logarithms to base e, the unit of information is the *nat* (natural unit).

To illustrate the difference between binary digit and binary unit, consider one random bit $X \in \{0, 1\}$. This random variable X follows a Bernoulli distribution with some parameter p. Its entropy, expressed in bits, is then

$$H(X) = p \log_2 \frac{1}{p} + (1-p) \log_2 \frac{1}{1-p} \quad \text{(in bits)} \tag{8}$$

which can take any value between 0 bit and 1 bit. The maximum value 1 bit is attained in the equiprobable case $p = 1/2$. Otherwise, the entropy of one bit is actually less than one bit.

The *Système Internationál d'unités* [61] recommends the use of the *shannon* (Sh) as the information unit in place of the *bit* to distinguish the amount of information from the quantity of data that may be used to represent this information. Thus according to the SI standard, $H(X)$ should actually be expressed in shannons. The entropy of one bit lies between 0 and 1 Sh.

11. H or Êta?

In information theory, following Shannon, the entropy is always denoted by the letter H. Where does this letter come from?

Ralph Hartley was perhaps Shannon's greatest influence, and he had already used the letter H – arguably his last name initial – as early as 1928 to denote the "amount of information" [25] with a formula identical to (7). Therefore, since Shannon generalized Hartley's measure of information, it seems logical that he would have adopted Hartley's letter H. In fact, Shannon did not at first use the name "entropy" for H but rather "uncertainty" [42]. All this seems to have nothing to do with the notion of entropy in physics.

Later Shannon adopted the term "entropy" [43] and mentioned that (5) is formally identical with Boltzmann's entropy in statistical mechanics, where $p(x)$ is the probability of a system being in a given cell x of its phase space. In fact, the very same letter H is used in Boltzmann's H-theorem to denote the *negative* continuous entropy (see § 15):

$$H = \int f \ln f \, \mathrm{d}^3 v, \tag{9}$$

where f denotes a distribution of particle velocities v. Boltzmann himself used the letter E at first [2], and it has been suggested that the first occurrence of the letter H in a paper by Burbury [4] was for "Heat" [30]. There is some indirect evidence, however, that in this context, H is in fact the capital greek letter H (Êta), the upper-case version of η, but the reason for which this choice was made is mysterious [28]. It does not seem to relate to the etymology of entropy, a term coined by Clausius [8] from the Greek $\varepsilon\nu\tau\rho o\pi\acute{\eta}$ ("inside transformation").

12. No One Knows What Entropy Really Is

Since the information-theoretic measure of information H is named *entropy* with reference to Boltzmann's entropy in statistical thermodynamics, the big question is: Is there a deep-lying connection between information theory and thermodynamics?

It is clear that the Shannon entropy is identical in form with previous expressions for entropy in statistical mechanics. The celebrated Boltzmann's entropy formula $S = k \log W$, where log denotes the natural logarithm and k is Boltzmann's constant equal to $1.3806485 \ldots 10^{-23}$ joules per kelvin, can be identified with (7) where $M = W$ is the number of microstates of a system in thermodynamic equilibrium. The integral version of entropy (with a minus sign) also appears in Boltzmann's first entropy formula $S = -\int \rho \ln \rho \, \mathrm{d}x$ where the probability distribution ρ represents the fraction of time spent by the system around a given point x of its space phase. Von Neumann's 1932 entropy formula $S = -\mathrm{Tr}(\hat{D} \log \hat{D})$ in quantum statistical mechanics [56] is also formally identical with (5) where $p(x)$ represent the eigenvalues of the density operator $\hat{D}$.

It is quite striking that such a strong formal analogy holds. Thus, although Shannon's information theory is certainly more mathematical than physical, any mathematical result derived in information theory could be useful when applied to physics with the appropriate interpretation.

Beyond the formal analogy, many physicists soon believed that a proper understanding of the second law of thermodynamics requires the notion of information. This idea can be traced back to Leó Szilárd [51] who attempted in the 1929 to solve Maxwell's demon problem by showing that an entropy decrease of $k \log 2$ per molecule is created by *intelligence* (the exactly informed Maxwell's demon). This was later recognized as the measure of *information* acquired by the demon, the term $k \log 2$ being identified with one "bit" of information. Szilárd was a personal friend of John von Neumann who derived his entropy formula a few years later. It is plausible that when von Neumann discovered Shannon's "information" formula, he immediately made the link with his entropy. In 1961, Shannon told Myron Tribus that von Neumann was the one who told him to call his new formula by the name 'entropy' in the early 1940s. According to Tribus, Shannon recalled:

> *"My greatest concern was what to call it. I thought of calling it 'information', but the word was overly used, so I decided to call it 'uncertainty'. When I discussed it with John von Neumann, he had a better idea. Von Neumann told me, 'You should call it entropy, for two reasons. In the first place your uncertainty function has been used in statistical mechanics under that name. In the second place, and more importantly, no one knows what entropy really is, so in a debate you will always have the advantage.'"* [53]

When asked twenty years later about this anecdote, however, Shannon did not remember von Neumann giving him such advice [47].

Norbert Wiener was also influenced by von Neumann who suggested to him the entropy formula $\int f(x) \log f(x)\, \mathrm{d}x$ as a "reasonable" measure of the amount of information associated with the curve $f(x)$ [59]. Shannon may have first come across the notion of entropy from Wiener, which was one of Shannon's teachers at MIT. Robert Fano, one of Shannon's colleagues at Bell Labs who worked on information theory in the early years, reported that when he was a PhD student at MIT, Wiener would at times enter his office, puffing at a cigar, saying "You know, information is entropy" [18]. Later the French-American physicist Léon Brillouin, building on Szilárd's and Shannon's works, coined the concept of *negentropy* to demonstrate the similarity between entropy and information [3].

Despite many attempts in the literature, it is still not clear why information theoretic principles should be necessary to understand statistical mechanics. Is there any physical evidence of a fundamental thermodynamic cost for the physical implementation of a informational computation, purely because of its logical properties? This is still a debated topic today [52]. As in the above Neumann-Shannon anecdote, no one knows what entropy really is.

13. How Does Entropy Arise Naturally?

Going back to mathematics, Shannon's entropy as a mathematical quantity arises in a fairly natural way. Shannon proposed the following line of reasoning [43]: Consider a long sequence of independent and identically distributed (i.i.d.) outcomes

$$\underline{x} = (x_1, x_2, \ldots, x_n). \tag{10}$$

To simplify, assume that the symbols x_i take a finite number of possibles values. Let $p(x)$ denote the probability that an outcome equals x. Thus each outcome follows the same probability distribution $p(x)$, of some random variable X.

By independence, the probability $p(\underline{x})$ of the long sequence $\underline{x}$ is given by the product

$$p(\underline{x}) = p(x_1)p(x_2)\cdots p(x_n). \tag{11}$$

Re-arrange factors according to the number $n(x)$ of x_i equal to x to obtain

$$p(\underline{x}) = \prod_x p(x)^{n(x)}, \tag{12}$$

where the product is over all possible outcome values x. Since n is taken very large, according to the law of large numbers, the empirical frequency of x can be identified to its probability:

$$\frac{n(x)}{n} \approx p(x). \tag{13}$$

Pugging this expression into (12) gives

$$p(\underline{x}) \approx \Big(\prod_x p(x)^{p(x)}\Big)^n = \exp\big(-nH(X)\big) \tag{14}$$

which exponentially decreases as $n \to +\infty$. The exponential decay $H(X) \geq 0$ is precisely given by Shannon's entropy (5). It is impressive to observe how this entropy arises "out of nowhere" in such a simple derivation.

Shannon's equation (14) is a fundamental result known as the *asymptotic equipartition property*: Essentially, this means that for very large (but fixed) n, the value of the probability of a given "typical" sequence $\underline{x} = (x_1, x_2, \ldots, x_n)$ is likely to be close to the constant $\exp(-nH(X))$. Moreover, any randomly chosen sequence is very likely to be "typical" (with probability arbitrarily close to one). As we have seen in the above derivation, this is essentially a consequence of the law of large numbers.

The fact that $p(\underline{x}) \approx \exp(-nH(X))$ for any typical sequence turns out to be very useful to solve the problem of information compression and other types of coding problems. This is perhaps the main mathematical justification of the usefulness of Shannon's entropy in science.

14. Shannon's Source Coding Theorem

The asymptotic equipartition property (14) for typical sequences is used to solve the *information compression* problem: How can we reliably encode a source of information with the smallest possible rate? This corresponds to the model of Figure 1 where the channel is noiseless and the information source is to be transmitted reliably at the destination while achieving the maximum possible compression.

Since non-typical sequences have a arbitrarily small probability, an arbitrary reliable compression is obtained by encoding typical sequences only. The resulting coding rate is computed from the number N of such typical sequences. Summing (14) over all the N typical sequences gives the total probability that a randomly chosen sequence is typical, which we know is arbitrarily close to one if the length n is taken sufficiently large:

$$1 \approx N \exp(-nH(X)). \tag{15}$$

This gives $N \approx \exp(nH(X))$ typical sequences. The resulting coding rate R is its logarithm per element in the sequence of length n:

$$R = \frac{\log N}{n} \approx H(X). \tag{16}$$

This is the celebrated *Shannon's first coding theorem* [43]: The minimal rate at which a source X can be encoded reliably is given by its entropy $H(X)$.

This important theorem provides the best possible performance of any data compression algorithm. In this context, Shannon's entropy receives a striking operational significance: It is the minimum rate of informational bits in a source of information.

15. Continuous Entropy

So far Shannon's entropy (5) was defined for discrete random variables. How is the concept generalized to continuous variables? An obvious way is to proceed by analogy. Definition (5) can be written as an expectation

$$H(X) = \mathbb{E} \log \frac{1}{p(X)}, \tag{17}$$

where $p(x)$ is the discrete distribution of X. When X follows a continuous distribution (pdf) $p(x)$, we may define its *continuous entropy* with formally the same formula:

$$h(X) = \mathbb{E} \log \frac{1}{p(X)} \tag{18}$$

which is

$$h(X) = \int p(x) \log \frac{1}{p(x)} \, \mathrm{d}x. \tag{19}$$

The discrete sum in (5) is simply replaced by an integral (a continuous sum).

Notice, however, that $p(x)$ does not refer to a probability anymore in (19), but to a probability *density*, which is not the same thing. For example, when the continuous random variable U is uniformly distributed over the interval (a, b), one has $p(u) = 1/(b - a)$ so that (18) becomes

$$h(U) = \mathbb{E} \log \frac{1}{p(U)} = \log(b - a). \tag{20}$$

While the discrete entropy is always nonnegative, the above continuous entropy expression becomes negative when the interval length is < 1. Moreover, taking the limit as the interval length tends to zero, we have

$$h(c) = -\infty \tag{21}$$

for any *deterministic* (constant) random variable $X = c$. This contrasts with the corresponding discrete entropy which is simply $H(c) = 0$.

Therefore, contrary to Shannon's entropy (5) for discrete variables, one cannot assign an "amount of information" to the continuous entropy $h(X)$ since it could be negative. Even though Shannon himself used the letter H for both discrete and continuous entropies [43], the capital H was soon degraded by information theorists to the lowercase letter h to indicate that the continuous entropy does not deserve the status of the genuine entropy H.

16. Change of Variable in the Entropy

In order to better understand why discrete and continuous entropies behave differently, consider $Y = T(X)$ with some invertible transformation T. If X is a discrete random variable, so is Y; the variables have different values but share the same probability distribution. Therefore, their discrete entropies coincide: $H(X) = H(Y)$. It is obvious, in this case, that X and Y should carry the same amount of information.

When X and $Y = T(X)$ are continuous random variables, however, their continuous entropies do *not* coincide. In fact, assuming T satisfies the requirements for an invertible change of variable (i.e., a diffeomorphism) with $\frac{\mathrm{d}y}{\mathrm{d}x} = T'(x) > 0$, the relation $p(x)\,\mathrm{d}x = \widetilde{p}(y)\,\mathrm{d}y$ gives

$$h(T(X) = h(Y) = \int \widetilde{p}(y) \log \frac{1}{\widetilde{p}(y)}\,\mathrm{d}y \tag{22}$$

$$= \int p(x) \log \frac{\mathrm{d}y/\,\mathrm{d}x}{p(x)}\,\mathrm{d}x \tag{23}$$

$$= \int p(x) \log \frac{1}{p(x)}\,\mathrm{d}x + \int p(x) T'(x)\,\mathrm{d}x, \tag{24}$$

hence the change of variable formula [43]:

$$h(T(X)) = h(X) + \mathbb{E} \log T'(X). \tag{25}$$

The difference $h(T(X)) - h(X)$ depends on the transformation T. For $T(x) = x + c$ where c is constant, we obtain

$$h(X + c) = h(X), \tag{26}$$

so the continuous entropy is invariant under shifts. For a linear transformation $T(x) = sx$ $(s > 0)$, however, we obtain the following *scaling property*:

$$h(sX) = h(X) + \log s. \tag{27}$$

Since $s > 0$ is arbitrary, the continuous entropy can take arbitrarily large positive or negative values, depending on the choice of s. For sufficiently small s (or sufficiently small variance), $h(X)$ becomes negative as we already have seen in the case of the uniform distribution (20).

17. Discrete vs. Continuous Entropy

Beyond the analogy between the two formulas, what is the precise relation between *discrete* entropy $H(X) = \sum p(x) \log \frac{1}{p(x)}$ and *continuous* entropy $h(X) = \int p(x) \log \frac{1}{p(x)}\,\mathrm{d}x$? To understand this, let us consider a *continuous* variable X with continuous density $p(x)$ and the corresponding *discrete* variable $[X]$ obtained by *quantizing* X with small quantization step δ. This means that we have a relation of the form

$$[X] = \delta \left\lfloor \frac{X}{\delta} \right\rfloor, \tag{28}$$

where $\lfloor \cdot \rfloor$ denotes the integer part. How $h(X)$ can be written in terms of $H([X])$? The integral (19) defining $h(X)$ can be approximated as a Riemann sum:

$$h(X) \approx \sum_k p(x_k) \log\Bigl(\frac{1}{p(x_k)}\Bigr) \delta x_k, \tag{29}$$

where $x_k = k\delta$, $\delta x_k = \delta$ and the approximation holds for small values of δ. Since the probability of a quantized value $[X] = k$ is

$$p(k) = \int_{k\delta}^{(k+1)\delta} p(x)\,\mathrm{d}x \approx p(x_k)\delta, \tag{30}$$

we obtain

$$h(X) \approx \sum_k p(k) \log\Bigl(\frac{\delta}{p(k)}\Bigr) = H([X]) - \log\frac{1}{\delta}. \tag{31}$$

This gives the desired relation between discrete and continuous entropies:

$$h(X) \approx H([X]) - \log\frac{1}{\delta}. \tag{32}$$

As $\delta \to 0$, $[X]$ converges to X but $H([X])$ does *not* converge to $h(X)$: In fact, $H([X]) - h(X) = \log(1/\delta)$ goes to $+\infty$. This confirms that discrete and continuous entropies behave very differently.

Interestingly, the entropy difference $\log(1/\delta)$ can be seen as the entropy of the difference $X - [X]$ (the quantization error). This is a consequence of (20) since such the quantization error approximately follows a uniform distribution in the interval $[0, \delta]$ of length δ. Thus (32) can be written as

$$h(X) \approx H([X]) - h(X - [X]). \tag{33}$$

From (33), the continuous entropy $h(X)$, also known as *differential* entropy, is obtained as the limit of a *difference* of two entropies. In particular, when X is deterministic, $H([X]) = 0$ and we recover that $h(X) = -\infty$ in this particular case. When X is a continuous random variable with finite differential entropy $h(X)$, since the limit of $H([X]) - h(X - [X])$ is finite as $\delta \to 0$ and $h(X - [X]) = \log(1/\delta) \to +\infty$, it follows that the discrete entropy $H([X])$ should actually diverge to $+\infty$:

$$\Bigl\{h(X) \text{ is finite}\Bigr\} \implies \Bigl\{H([X]) \to +\infty\Bigr\}. \tag{34}$$

This is not surprising in light of Shannon's first coding theorem (16): An arbitrarily fine quantization of a continuous random variable requires a arbitrarily high precision and therefore, an infinite coding rate.

18. Most Beautiful Equation

It has long been said that the most beautiful mathematical equation is Euler's identity $e^{i\pi}+1=0$, because it combines the most important constants in mathematics like π and e together. Here's another one that should perhaps be considered equally beautiful. Let $X^* \sim \mathcal{N}(0,1)$ by the standard normal, with density

$$p(x^*) = \frac{1}{\sqrt{2\pi}} e^{-\frac{x^{*2}}{2}}. \tag{35}$$

This of course is a fundamental distribution in mathematics and in physics, the limit of the well-known central limit theorem. Its entropy is easily computed from (18) as

$$h(X^*) = \mathbb{E}\log\bigl(\sqrt{2\pi}e^{X^{*2}/2}\bigr) \tag{36}$$

$$= \log\sqrt{2\pi} + \log e \cdot \mathbb{E}(X^{*2})/2. \tag{37}$$

Here $\mathbb{E}(X^{*2})=1$ since the standard normal has zero mean and unit variance. We obtain

$$\boxed{h(X^*) = \log\sqrt{2\pi e}}, \tag{38}$$

a lovely formula that combines the three most important real constants in mathematics: $\sqrt{2}$ (diagonal of a unit square), π (circumference of a circle with unit diameter), and e (base of natural logarithms).

The more general case where $X^* \sim \mathcal{N}(\mu,\sigma^2)$ follows a Gaussian distribution with mean μ and variance σ^2 is obtained by multiplying the standard variable by σ and adding μ. From (26) and (27) we obtain

$$h(X^*) = \log\sqrt{2\pi e} + \log\sigma = \frac{1}{2}\log(2\pi e\sigma^2). \tag{39}$$

19. Entropy Power

Shannon advocated the use of the *entropy power* rather than the entropy in the continuous case [43]. Loosely speaking, the entropy power is defined as the power of the noise having the same entropy. Here the noise considered is the most common type of noise encountered in engineering, sometimes known as "thermal noise", and modeled mathematically as a zero-mean Gaussian random variable X^*. The (average) noise power N^* is the mean squared value $\mathbb{E}(X^{*2})$ which equals the variance of X^*. By (39) its entropy is

$$h(X^*) = \frac{1}{2}\log(2\pi e N^*) \tag{40}$$

so that

$$N^* = \frac{\exp\bigl(2h(X^*)\bigr)}{2\pi e}. \tag{41}$$

The entropy power $N(X)$ of a continuous random variable X is, therefore, the power N^* of the noise X^* having the same entropy $h(X^*) = h(X)$. This gives

$$N(X) = \frac{\exp\bigl(2h(X)\bigr)}{2\pi e}. \tag{42}$$

Interestingly, it turns out that the "entropy *power*" is essentially a constant raised to the *power* of the (continuous) entropy. Thus, the physicist's view of entropy power uses the notion of power in physics while the mathematician's view refers to the notion of power in the mathematical operation of exponentiation.

When X is itself zero-mean Gaussian, its entropy power equals its actual power $\mathbb{E}(X^2)$. In general, X is not necessarily Gaussian, but the entropy power still satisfies some properties that one would expect for a power: It is a positive quantity, with the following scaling property:

$$N(aX) = a^2 N(X) \tag{43}$$

which is an immediate consequence of (27).

20. A Fundamental Information Inequality

A fundamental inequality, first derived by Gibbs in the 19th century [23], is sometimes known as the *information inequality* [9, Thm. 2.6.3]: For any random variable X with distribution $p(x)$,

$$\mathbb{E} \log \frac{1}{p(X)} \leq \mathbb{E} \log \frac{1}{q(X)}, \tag{44}$$

where the expectation is taken with respect to p, and where $q(x)$ is any other probability distribution. Equality holds if and only if distributions p and q coincide. The left-hand side of (44) is the discrete or continuous entropy, depending on whether the variable X is discrete or continuous. Thus

$$H(X) = \sum_x p(x) \log \frac{1}{p(x)} \leq \sum_x p(x) \log \frac{1}{q(x)} \tag{45}$$

when X is discrete with probability distribution $p(x)$ and

$$h(X) = \int p(x) \log \frac{1}{p(x)} \,\mathrm{d}x \leq \int p(x) \log \frac{1}{q(x)} \,\mathrm{d}x \tag{46}$$

when X is continuous with probability density $p(x)$. Notice that the right-hand side is always identical to the left-hand side except for the distribution inside de logarithm.

Gibbs' inequality (44) is an easy consequence of the concavity of the logarithm. By Jensen's inequality, the difference between the two sides of (44) is

$$\mathbb{E} \log \frac{q(X)}{p(X)} \leq \log \mathbb{E} \frac{q(X)}{p(X)} = \log 1 = 0. \tag{47}$$

Indeed, $\mathbb{E}\,\frac{q(X)}{p(X)} = \sum_x \frac{q(x)}{p(x)} p(x) = \sum_x q(x) = 1$ in the discrete case, and $\mathbb{E}\,\frac{q(X)}{p(X)} = \int \frac{q(x)}{p(x)} p(x)\,\mathrm{d}x = \int q(x)\,\mathrm{d}x = 1$ in the continuous case. Because the logarithm is strictly concave, equality in Jensen's inequality holds if and only if $q(x)/p(x)$ is constant, which implies that the two distributions $p(x)$ and $q(x)$ coincide.

The fundamental information inequality (44) is perhaps the most important inequality in information theory because, as seen below, every classical information-theoretic inequality can be easily derived from it.

21. The MaxEnt Principle

The maximum entropy (MaxEnt) principle first arose in statistical mechanics, where it was shown that the maximum entropy distribution of velocities in a gas under the temperature constraint is the Maxwell-Boltzmann distribution. The principle has been later advocated by Edwin Jaynes for use in a general context as an attempt to base the laws of thermodynamics on information theory [29]. His "MaxEnt school" uses Bayesian methods and has been sharply criticized by the orthodox "frequentist" school [14]. In an attack against the MaxEnt interpretation, French mathematician Benoît Mandelbrot once said: "Everyone knows that Shannon's derivation is in error"[54]. Of course, as we now show, Shannon's mathematical derivation is mathematically correct. Only physical misinterpretations of his calculations could perhaps be questionable.

Consider the following general maximum entropy problem: Maximize the (discrete or continuous) entropy over all random variables satisfying a constraint of the form $\mathbb{E}\{w(X)\} = \alpha$, where $w(x)$ is a some given weight function. A classical approach to solving the problem would use the Lagrangian method, but a much simpler derivation is based on Gibbs' inequality (44) as follows.

Consider the "exponential" probability distribution

$$q(x) = \frac{e^{-\lambda w(x)}}{Z(\lambda)}, \tag{48}$$

where $Z(\lambda)$ is a normalizing factor, known in physics as the canonical partition function, and λ is chosen so as to meet the constraint $\mathbb{E}\{w(X)\} = \alpha$. Plugging (48) into (44) gives an upper bound on the discrete or continuous entropy:

$$\begin{aligned} H(X) \text{ or } h(X) \quad &\leq \mathbb{E} \log\bigl(Z(\lambda) e^{\lambda w(X)}\bigr) && (49)\\ &= \log Z(\lambda) + (\log e)\lambda\, \mathbb{E}\{w(X)\} && (50)\\ &= \log Z(\lambda) + \alpha\lambda \log e. && (51)\end{aligned}$$

The entropy's upper bound has now become constant, independent of the probability distribution $p(x)$ of X. Since equality (44) holds if and only if $p(x)$ and $q(x)$ coincide, the upper bound (51) is attained precisely when $p(x)$ is given by (48). Therefore, $\log Z(\lambda) + \alpha\lambda \log e$ is in fact the desired value of the maximum entropy. The above method can be easily generalized in the same manner to more than one constraint.

This general result, sometimes known as the *Shannon bound*, can be applied to many important problems. First, what is the maximum entropy of a discrete random variable that can take at most M values? Set $w(x) = 0$, $\alpha = 0$ so that $Z(\lambda) = M$ where the actual value of λ is of no importance. Then

$$\max H(X) = \log M \tag{52}$$

attained for a uniform distribution. Thus (7) is the maximum uncertainty, when all outcomes are equally probable: One event cannot be expected in preference to another. This is the classical assumption in the absence of any prior knowledge.

Similarly, what is the maximum entropy of a continuous random variable having values in a finite-length interval $[a, b]$? Again set $w(x) = 0$, $\alpha = 0$ so that $Z(\lambda) = b - a$, the interval length. Then

$$\max_{X \in [a,b]} h(X) = \log(b - a). \tag{53}$$

Thus (20) is the maximum entropy, attained for a uniform distribution on $[a, b]$.

More interestingly, what is the maximum entropy of a continuous variable with fixed mean μ and variance σ^2? Set $w(x) = (x - \mu)^2$, then $\alpha = \sigma^2$, $\lambda = 1/2\sigma^2$, $Z(\lambda) = \sqrt{2\pi/\sigma^2}$, hence the maximum entropy is $\log \sqrt{2\pi\sigma^2} + (1/2) \log e$:

$$\max_{\mathsf{Var}\, X = \sigma^2} h(X) = \frac{1}{2} \log(2\pi e \sigma^2) \tag{54}$$

attained for a normal $\mathcal{N}(\mu, \sigma^2)$ distribution. In other words, (39) is the maximum entropy for fixed variance. When X is zero mean, $\sigma^2 = \mathbb{E}(X^2)$ is its power, hence the entropy power (42) cannot exceed the actual power, which is attained if and only if X is a Gaussian random variable.

The fact that the Gaussian (normal) distribution maximizes the entropy for fixed first and second moments is of paramount importance in many engineering methods, such as Burg's spectral estimation method [5].

22. Relative Entropy or Divergence

The fundamental information inequality (44) gives rise to a new informational measure which is in many respects even more fundamental than the entropy itself. Let X be distributed according to the distribution $p(x)$ and let X^* be distributed according to another distribution $q(x)$. Then the difference between the two sides of (44) is the *relative entropy*

$$D(X, X^*) = \mathbb{E} \log \frac{p(X)}{q(X)} \geq 0, \tag{55}$$

often noted $D(p, q)$ to stress the dependence on the two distributions[1]. This is also known as the Kullback-Leibler *divergence* $D(p, q)$ between the two distributions p and q [33].

---

[1]It has now become common practice for information theorists to adopt the notation $D(p\|q)$ with a double vertical bar. The origin of such an unusual notation seems obscure.

In contrast to Shannon's entropy, the relative entropy is positive in both cases of discrete or continuous distributions:

$$D(X, X^*) = \sum_x p(x) \log \frac{p(x)}{q(x)} \geq 0 \tag{56}$$

for discrete probability distributions p, q and

$$D(X, X^*) = \int p(x) \log \frac{p(x)}{q(x)} \, \mathrm{d}x \geq 0 \tag{57}$$

for probability density functions p, q. In addition, the relative entropy $D(X, X^*) = D(p, q)$ vanishes if and only if equality holds in (44), that is, when p and q coincide. Therefore, $D(X, X^*) = D(p, q)$ can be seen as a measure of "informational distance" relative to the two distributions p and q. Notice, however, that the above expressions are *not* symmetrical in (p, q).

Furthermore, if we consider continuous random variables X, X^* and quantize them as in (28) with small quantization step δ to obtain discrete random variables $[X], [X^*]$, then the $\log \frac{1}{\delta}$ term present in § 17 cancels out on both sides of (46) and we obtain

$$D([X], [X^*]) \to D(X, X^*) \qquad \text{as } \delta \to 0. \tag{58}$$

Thus, as the discrete random variables $[X], [X^*]$ converge to the continuous ones X, X^*, their relative entropy $D([X], [X^*])$ similarly converge to $D(X, X^*)$. This important feature of divergence allows one to deduce properties for continuous variables from similar properties derived for discrete variables.

Finally, in the MaxEnt principle described in § 21, letting $q(x)$ (the distribution of X^*) be the entropy-maximizing distribution (48), we see that the right-hand side of Gibbs' inequality (44) equals the maximum entropy of X^*. Therefore, in this case, the divergence is simply the difference between the entropy and its maximum value:

$$D(X, X^*) = \begin{cases} H(X^*) - H(X) & \text{in the discrete case,} \\ h(X^*) - h(X) & \text{in the continuous case.} \end{cases} \tag{59}$$

For example, for M-ary variables, $D(X, X^*) = H(X^*) - H(X) = \log M - H(X)$ can be seen as a measure of redundancy: It is the amount of rate reduction performed by an optimal coding scheme according to Shannon's source coding theorem (§ 14). When X and X^* are continuous variables with the same variance σ^2, X^* is normally distributed and $D(X, X^*) = h(X^*) - h(X)$ represents the "non-Gaussianity" of the random variable X, which vanishes if and only if X is Gaussian.

23. Generalized Entropies and Divergences

There exist numerous generalizations of Shannon's entropy and relative entropy. In 1960, Alfréd Rényi looked for the most general definition of information measures that would preserve the additivity of independent events [37]. The Rényi entropy is defined for discrete random variables as

$$H_\alpha(X) = \frac{1}{1-\alpha}\log\sum_x p(x)^\alpha \tag{60}$$

and the continuous version is accordingly $h_\alpha(X) = \frac{1}{1-\alpha}\log\int p(x)^\alpha\,\mathrm{d}x$. One recovers Shannon's entropy by letting $\alpha \to 1$. The most interesting special cases are $\alpha = 0$ (the max-entropy), $\alpha = \infty$ (the min-entropy) and $\alpha = 2$ (the collision entropy). There is also a Rényi α-divergence $D_\alpha(p,q) = \frac{1}{\alpha-1}\log\sum_x p(x)^\alpha/q(x)^{\alpha-1}$. Rényi entropies have found many applications such as source coding, hypothesis testing, channel coding, guessing, quantum information theory, and computer science.

The Tsallis entropy, first introduced by Havrda and Charvát [26], was proposed as a basis for generalizing the standard Boltzmann-Gibbs statistical mechanics [55]. Since then its physical relevance has been debated [7]. It is defined as

$$\frac{1}{\alpha-1}\bigl(1-\sum_x p(x)^\alpha\bigr) = \frac{1-\exp\bigl((1-\alpha)H_\alpha(X)\bigr)}{\alpha-1} \tag{61}$$

with the continuous version $\frac{1}{\alpha-1}\bigl(1-\int p(x)^\alpha\,\mathrm{d}x\bigr)$. Again Shannon's entropy is recovered by letting $\alpha \to 1$.

All these entropies and relative entropies have been further extensively generalized as *f-divergences* [11] for some convex function f. Instances of f-divergences are: relative entropy (Kullback-Leibler divergence), Rényi and Tsallis divergences, the Hellinger distance, the Jensen-Shannon divergence, Vajda divergences including the total variation distance and the Pearson (χ^2) divergence, etc. There is abundant literature on such generalized concepts and their applications in signal processing, statistics and information theory.

24. How Does Relative Entropy Arise Naturally?

Similarly as in § 13, the relative entropy receives a useful operational justification. Going back to the expression (12) for the probability of a sequence $\underline{x}$ of n independent outcomes, and letting

$$q(x) = \frac{n(x)}{n} \tag{62}$$

be the empirical probability of the sequence $\underline{x}$ (also referred to as its *type*), we can rewrite (14) as an *exact* expression

$$p(\underline{x}) = \Bigl(\prod_x p(x)^{q(x)}\Bigr)^n = \exp\bigl(-nH(X,X^*)\bigr), \tag{63}$$

where $H(X,X^*)$ is the so-called *cross-entropy*

$$H(X,X^*) = \sum_x q(x)\log\frac{1}{p(x)}. \tag{64}$$

Since a typical sequence is characterized by its type q, the probability that a randomly chosen sequence is typical is exactly $N\exp(-nH(X,X^*))$, where N is the number of typical sequences. Thus in particular $N\exp(-nH(X,X^*)) \leq 1$ for any choice of p, and in particular for $p = q$ we have $N\exp(-nH(X^*,X^*)) \leq 1$, that is, $N \leq \exp(-nH(X^*))$. Since

$$H(X^*,X^*) - H(X^*) = \sum_x q(x) \log \frac{1}{p(x)} - \sum_x q(x) \log \frac{1}{q(x)} = D(q,p), \quad (65)$$

the probability that a randomly chosen sequence is typical (according to the actual probability distribution p) is bounded by

$$N\exp(-nH(X,X^*)) \leq \exp(-nD(q,p)), \quad (66)$$

where $D(q,p) = D(X^*,X) \geq 0$ is the relative entropy or divergence. Therefore, if $q(x)$ diverges from $p(x)$, the exponent $D(q,p)$ is strictly positive and the probability (66) can be made exponentially small.

Juste like the asymptotic equipartition property (14) is used to solve the *information compression* problem (Shannon's source coding theorem in § 14), the above asymptotic "large deviation" bound (66) will be used in § 32 to solve the *information transmission* problem (Shannon's channel coding theorem).

25. Chernoff Information

Derivations similar to the above in the preceding section (§ 24) form the basis of the *method of types*, a powerful technique in large deviations theory. More generally, there is a strong relationship between information theory and statistics, and the Kullback-Leibler divergence (55) has become a fundamental tool for solving many problems in statistics.

For example, in the problem of testing hypotheses, the Kullback-Leibler divergence is used to derive the best possible error exponents for tests to decide between two alternative i.i.d. distributions p and q. In a Bayesian approach where we assign prior probabilities to both hypotheses, the exponent of the overall probability error is given by

$$D(X_\lambda, X) = D(X_\lambda, X^*), \quad (67)$$

where X follows p, X^* follows q, and X_λ follows a distribution r_λ proportional to $p^\lambda q^{1-\lambda}$. Here $\lambda \in [0,1]$ is chosen such that equality (67) holds, which gives the maximum error exponent.

The common value (67) is known as the *Chernoff information* $C(X,X^*)$. Just as for the Kullback-Leibler divergence, it was derived in the early 1950s [6]. An easy calculation shows that

$$C(X,X^*) = \max_{\lambda}\big(\lambda D(X_\lambda, X) + (1-\lambda) D(X_\lambda, X^*)\big) \tag{68}$$

$$= \max_{\lambda} \mathbb{E} \log \frac{r_\lambda(X_\lambda)}{p^\lambda(X_\lambda) q^{1-\lambda}(X_\lambda)} \tag{69}$$

$$= -\min_{\lambda} \log \sum_x p^\lambda(x) q^{1-\lambda}(x). \tag{70}$$

Such an information measure is symmetric in (p,q), positive and vanishes if and only if the two distributions p and q coincide. Today, Chernoff information plays an important role as a statistical distance for various data processing applications.

26. Fisher Information

In statistical parametric estimation, the concept of *information* was already explored by Ronald Fisher in the 1920s [20], following an early work of Edgeworth [15] forty years before Shannon. Loosely speaking, Fisher's information measures the amount of information about a parameter θ in an observed random variable X, where X is modeled by a probability density $p_\theta(x)$ that depends on θ.

To understand the significance of the Fisher information, consider an *estimator* of θ, that is, some function $\hat{\theta}(X)$ of the observed random variable X that is used to estimate the value of θ. An optimal estimator would minimize the mean-squared error (MSE), given by

$$\text{MSE} = \mathbb{E}\big((\hat{\theta}(X) - \theta)^2\big) = \int (\hat{\theta}(x) - \theta)^2 \, p_\theta(x) \, \mathrm{d}x. \tag{71}$$

Suppose, for simplicity, that the estimator is *unbiased*, i.e., its bias is zero for any value of θ:

$$\text{Bias} = \mathbb{E}(\hat{\theta}(X) - \theta) = \int (\hat{\theta}(x) - \theta) \, p_\theta(x) \, \mathrm{d}x = 0. \tag{72}$$

Taking the derivative with respect to θ, we obtain

$$1 = \int (\hat{\theta}(x) - \theta) \frac{\partial p_\theta}{\partial \theta}(x) \, \mathrm{d}x \tag{73}$$

$$= \int (\hat{\theta}(x) - \theta) S_\theta(x) p_\theta(x) \, \mathrm{d}x \tag{74}$$

$$= \mathbb{E}\big((\hat{\theta}(X) - \theta) S_\theta(X)\big), \tag{75}$$

where

$$S_\theta(x) = \frac{\frac{\partial p_\theta}{\partial \theta}(x)}{p_\theta(x)} = \frac{\partial \log p_\theta}{\partial \theta}(x) \tag{76}$$

is known as the *score* or *informant*, the derivative of the log-likelihood with respect to θ. Now, by the Cauchy-Schwarz inequality,

$$1 = \big\{ \mathbb{E}\big((\hat{\theta}(X) - \theta) S_\theta(X)\big) \big\}^2 \leq \mathbb{E}\big((\hat{\theta}(X) - \theta)^2\big) \cdot \mathbb{E}\big(S_\theta(X)^2\big), \tag{77}$$

where

$$J_\theta(X) = \mathbb{E}\big(S_\theta(X)^2\big) = \mathbb{E}\Big(\Big(\frac{\partial \log p_\theta}{\partial \theta}(X)\Big)^2\Big) \tag{78}$$

is the *Fisher information.* The above inequality now writes

$$\text{MSE} \geq \frac{1}{J_\theta(X)}. \tag{79}$$

This is the celebrated *Cramér-Rao inequality* derived by Fréchet, Darmois, Rao and Cramér in the early 1940s [21, 12, 36, 10]. This inequality states that a universal lower bound on the mean-squared error of any unbiased estimator is given by the reciprocal of the Fisher information. In other words, the larger amount of information $J_\theta(X)$ about θ, the more reliable its (unbiased) estimation can be.

Despite appearances, there is a strong relationship between Fisher's and Shannon's concepts of information. In fact, Fisher's information can be expressed in terms of the relative entropy (divergence)

$$D(p_\theta, p_{\theta'}) = \mathbb{E}\log\frac{p_\theta(X)}{p_{\theta'}(X)} = \int p_\theta(x)\log\frac{p_\theta(x)}{p_{\theta'}(x)}\,\mathrm{d}x. \tag{80}$$

By the fundamental information inequality (44), we know that $D(p_\theta, p_{\theta'})$ is positive and vanishes when $\theta' = \theta$ (since in this case the two distributions p_θ and $p_{\theta'}$ coincide). Therefore, the derivative with respect to θ' vanishes at $\theta' = \theta$, and the second derivative is positive and represents its curvature at $\theta' = \theta$.

For the first derivative, we have

$$\frac{\partial}{\partial \theta'} D(p_\theta, p_{\theta'})\Big|_{\theta'=\theta} = -\,\mathbb{E}\,\frac{\partial}{\partial \theta'}\log p_{\theta'}(X)\Big|_{\theta'=\theta} = -\,\mathbb{E}\big(S_\theta(X)\big) = 0 \tag{81}$$

which simply means that the score has zero mean – hence the Fisher information (78) also equals the *variance* of the score. That $\mathbb{E}\big(S_\theta(X)\big) = 0$ can easily be checked directly since $\int p_\theta(x)\frac{\frac{\partial p_\theta(x)}{\partial\theta}}{p_\theta(x)}\,\mathrm{d}x = \int \frac{\partial p_\theta(x)}{\partial \theta}\,\mathrm{d}x = \frac{\partial}{\partial\theta}\int p_\theta(x)\,\mathrm{d}x = \frac{\partial 1}{\partial \theta} = 0$.

For the second derivative, we have

$$\frac{\partial^2}{\partial \theta'^2} D(p_\theta, p_{\theta'})\Big|_{\theta'=\theta} = -\,\mathbb{E}\,\frac{\partial^2}{\partial \theta^2}\log p_\theta(X) \tag{82}$$

which is the expected value of $-\frac{\partial^2}{\partial\theta^2}\log p_\theta(X)$, sometimes referred to as the *observed information.* Expanding $\frac{\partial^2}{\partial\theta^2}\log p_\theta(x) = \frac{\partial}{\partial\theta}\frac{\frac{\partial p_\theta}{\partial\theta}(x)}{p_\theta(x)} = \frac{\frac{\partial^2 p_\theta}{\partial\theta^2}(x)}{p_\theta(x)} - \big(\frac{\frac{\partial p_\theta}{\partial\theta}(x)}{p_\theta(x)}\big)^2 = \frac{\frac{\partial^2 p_\theta}{\partial\theta^2}(x)}{p_\theta(x)} - S_\theta(x)^2$, one finds that $\mathbb{E}\,\frac{\frac{\partial^2 p_\theta}{\partial\theta^2}(X)}{p_\theta(X)} = \int p_\theta(x)\frac{\frac{\partial^2 p_\theta(x)}{\partial\theta^2}}{p_\theta(x)}\,\mathrm{d}x = \int\frac{\partial^2 p_\theta(x)}{\partial\theta^2}\,\mathrm{d}x = \frac{\partial^2}{\partial\theta^2}\int p_\theta(x)\,\mathrm{d}x = \frac{\partial^2 1}{\partial\theta^2} = 0$. Therefore,

$$\frac{\partial^2}{\partial \theta'^2} D(p_\theta, p_{\theta'})\Big|_{\theta'=\theta} = J_\theta(X). \tag{83}$$

Thus, the Fisher information also equals the expected "observed information" (82) and can be identified to the *curvature* of the relative entropy. In fact, the second-order Taylor expansion of relative entropy about θ is

$$D(p_\theta, p_{\theta'}) = \frac{1}{2} J_\theta(X) \cdot (\theta' - \theta)^2 + o(\theta' - \theta)^2. \tag{84}$$

Thus, the more information about θ, the more "sharply peaked" is the relative entropy about its minimum at θ. This means that θ is all more sharply localized as its Fisher information is large.

27. Kolmogorov Information

We have seen in § 9 that when X is a random variable with probability distribution $p(x)$, the amount of information associated to the event $X = x$ can be defined as $\log \frac{1}{p(x)}$. From Shannon's source coding theorem (§ 14), $\log \frac{1}{p(x)}$ represents the minimal bit length required to describe x by an optimal code. Thus, the original approach of Shannon is to base information theory on *probability theory*, which incidentally was axiomatized as a rigorous mathematical theory by Andreï Kolmogorov in the 1930s.

Kolmogorov was an ardent supporter of Shannon's information theory in the 1950s and 1960s. He went further than Shannon by defining the algorithmic *complexity* of x as the length of the shortest binary computer program that describes x. Kolmogorov proved that not only his definition of complexity is essentially computer independent, but also that the average algorithmic complexity of a random variable is roughly equal to its entropy. In this way, the Kolmogorov complexity extends Shannon's entropy while dispensing with the notion of probability distribution. In a summary of his work on complexity theory, Kolmogorov wrote:

> *"Information theory must precede probability theory, and not be based on it. By the very essence of this discipline, the foundations of information theory have a finite combinatorial character."* [32]

The concept of Kolmogorov's information or complexity is perhaps more philosophical than practical, closely related to Turing machines, Church's thesis, universal codes, the Occam's razor principle, and Chaitin's mystical number Ω – a well-known "philosopher's stone". The reader is referred to [9, Chap. 14] for a more detailed introduction.

28. Shannon's Mutual Information

As we already have noted, Claude Shannon used the term 'communication theory' in his seminal 1948 work, not 'information theory'. However, his most important results rely on the notion of transmitted information over a communication channel. This was soon formalized by Robert Fano who coined the term 'mutual information'. Fano recalled:

> *I didn't like the term 'information theory'. Claude didn't like it either. You see, the term 'information theory' suggests that it is a theory about information – but it's not. It's the transmission of information, not information. Lots of people just didn't understand this. I coined the term 'mutual information' to avoid such nonsense: making the point that information is always about something. It is information provided by something, about something.* [19]

Thanks to the notion of divergence $D(p,q)$ (§ 22), Shannon's mutual information can be easily defined as a measure of mutual dependence between two random variables X and Y. Let $p(x,y)$ be the joint distribution of X and Y. If X and Y were independent, the joint distribution would equal the product of marginals: $p(x,y) = p(x)p(y)$. In general, however, the two distributions $p(x,y)$ and $q(x,y) = p(x)p(y)$ do not coincide. The *mutual information* $I(X;Y)$ is simply the divergence $D(p,q) = \mathbb{E}\log\frac{p(X,Y)}{q(X,Y)}$:

$$I(X;Y) = \mathbb{E}\log\frac{p(X,Y)}{p(X)p(Y)}. \tag{85}$$

It is a measure of mutual dependence as expected: By the fundamental information inequality (44), $I(X;Y) \geq 0$ with equality $I(X;Y) = 0$ if and only if $p = q$, that is, X and Y are independent.

The same definition and properties hold for both discrete or continuous random variables, thanks to the property (58) of divergence. Thus, if $[X]$ and $[Y]$ are quantized versions of continuous variables X, Y then $I([X];[Y]) \to I(X;Y)$ as the quantization step $\delta \to 0$.

Notice that although divergence $D(p,q)$ is not symmetric in (p,q), mutual information *is* symmetric in (X,Y): $I(X;Y) = I(Y;X)$, hence the term *mutual.* It was found convenient by information theorists to use the semi colon ';' as the argument separator in the mutual information with lower precedence over the comma ',' e.g., to make the distinction between mutual informations $I(X,Y;Z)$ (between (X,Y) and Z) and $I(X;Y,Z)$ (between X and (Y,Z)).

As usual for informational measures, mutual information $I(X;Y)$ does not actually depend on the real values taken by the variables X, Y, but only on their probability distributions. Thus, mutual information is well defined even for categorical variables. This can be seen as an advantage over other dependence measures like linear (or nonlinear) correlation.

As in Fano's quote above, mutual information $I(X;Y)$ can be interpreted as a measure of information *provided by* Y *about* X. To see this, rewrite (85) as

$$I(X;Y) = \mathbb{E}\log\frac{p(X|Y)}{p(X)}, \tag{86}$$

where $p(x|y) = p(x,y)/p(y)$ is the conditional distribution of X *knowing* $Y = y$. The (unconditional) distribution $p(x)$ of X (*not* knowing Y) is affected by the knowledge of Y and the corresponding average relative entropy is precisely the mutual information (86). By symmetry, $I(X;Y)$ is also a measure of information *provided by X about Y*.

The concept of mutual information has been generalized to more than two variables although the corresponding multivariate mutual information can sometimes be negative [17].

29. Conditional Entropy or Equivocation

As seen below, mutual information (86) is a central concept in Shannon's information theory. It can be easily related to the concept of entropy by simply rewriting (86) as

$$I(X;Y) = \mathbb{E}\log\frac{1}{p(X)} - \mathbb{E}\log\frac{1}{p(X|Y)}. \tag{87}$$

Thus mutual information is the difference between Shannon's entropy of X and a "conditional" entropy of X given Y:

$$I(X;Y) = \begin{cases} H(X) - H(X|Y) & \text{in the discrete case,} \\ h(X) - h(X|Y) & \text{in the continuous case,} \end{cases} \tag{88}$$

where the *conditional entropy*, also known as *equivocation*, is defined by

$$H(X|Y) = \sum_x\sum_y p(x,y)\log\frac{1}{p(x|y)} \tag{89}$$

in the discrete case, and

$$h(X|Y) = \iint p(x,y)\log\frac{1}{p(x|y)}\,\mathrm{d}x\,\mathrm{d}y \tag{90}$$

in the continuous case. Notice that in contrast to the entropy of one variable, the above definitions of conditional entropy involve averaging over both variables X and Y. By symmetry of mutual information the variables X and Y can be interchanged in the above expressions. There is also a notion of *conditional mutual information*, e.g., $I(X;Y|Z) = H(X|Z) - H(X|Y,Z)$.

30. Knowledge Reduces Uncertainty – Mixing Increases Entropy

The conditional entropy can be written as an average value of entropies, e.g.,

$$H(X|Y) = \sum_y p(y)\sum_x p(x|y)\log\frac{1}{p(x|y)} = \sum_y p(y)H(X|Y=y). \tag{91}$$

Thus, for discrete variables, while $H(X)$ measures the uncertainty about X, $H(X|Y)$ is a measure of the average uncertainty about X *when Y is known*. By (88), $I(X;Y) = H(X) - H(X|Y) \geq 0$, hence *knowledge reduces uncertainty* (on average):

$$H(X|Y) \leq H(X). \tag{92}$$

The difference between the two uncertainties is precisely $I(X;Y)$, the amount of information provided by Y about X.

Similarly for continuous variables, since $I(X;Y) = h(X) - h(X|Y) \geq 0$, we still observe that *conditioning reduces entropy*:

$$h(X|Y) \leq h(X), \tag{93}$$

even though we have seen that these differential entropies cannot be interpreted as uncertainty measures.

As an interesting application of (92) consider two systems with probability distributions p_1 and p_2 and their linear mixture $p_\lambda = \lambda_1 p_1 + \lambda_2 p_2$ where $\lambda_1 \geq 0$ and $\lambda_2 \geq 0$ are such that $\lambda_1 + \lambda_2 = 1$. If X_1 follows p_1 and X_2 follows p_2, then $p_\lambda = \lambda_1 p_1 + \lambda_2 p_2$ can be seen as the distribution of the random variable X_λ, where $\lambda \in \{1, 2\}$ is itself random with respective probabilities λ_1, λ_2. Then by (92), the entropy of the mixture satisfies

$$H(X_\lambda) \geq H(X_\lambda|\lambda) = \lambda_1 H(X_1) + \lambda_2 H(X_2). \tag{94}$$

In other words, *mixing increases entropy*: the entropy of the mixture is not less than the corresponding mixture of entropies. This can also be seen as a *concavity* property of entropy (with respect to the probability distribution), a classical statement in information theory.

It is quite fascinating to see how such exciting expressions such as "knowledge reduces uncertainty" or "mixing increases entropy" receive a rigorous treatment in information theory. This perhaps explains the extraordinary wave of popularity that Shannon's theory has experienced in the past. More suggestive results are derived in the next section.

31. A Suggestive Venn Diagram

The relationship between entropies, conditional entropies and mutual information is summarized for discrete variables in the Venn diagram of Figure 2. Using the diagram, we recover the relations $I(X;Y) = H(X) - H(X|Y) = H(Y) - H(Y|X)$. In addition, many useful properties of information measures can be derived from it.

If, for example, $X = Y$ (or if X and Y are in bijection) then the two uncertainty sets coincide in Figure 2, $H(X|Y) = H(Y|X) = 0$, and

$$H(X) = I(X;X). \tag{95}$$

This means that *self-information is entropy*: $H(X)$ can be seen as the measure of information provided by X about X itself. In particular, we recover that $H(X) \geq 0$, with equality $H(X) = 0$ if and only if X is independent of itself (!) which simply means that $p(x) = 0$ or 1, i.e., X is deterministic or *certain*. This confirms the intuition that $H(X)$ is measure of randomness or uncertainty. For a continuous random variable, we would have $I(X;X) = H(X) = +\infty$ as explained in § 17.

For independent variables X and Y, $I(X;Y) = 0$ and the two sets in Figure 2 are disjoint. In this case the joint entropy is $H(X,Y) = H(X)+H(Y)$, the individual

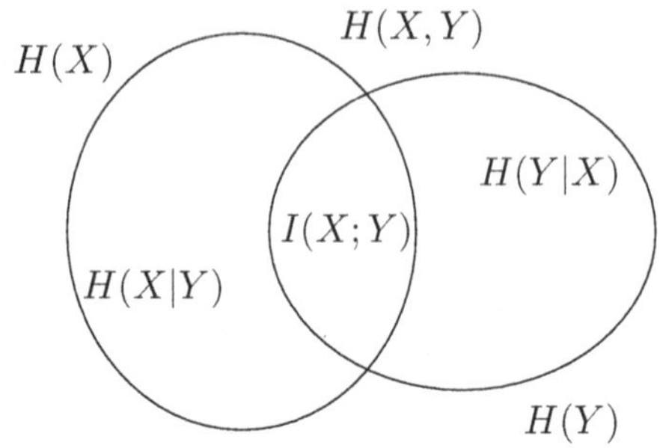

FIGURE 2. Venn diagram illustrating relationships among Shannon's measures of information. The mutual information $I(X;Y)$ corresponds to the intersection of the two "uncertainty sets", while the *joint entropy* $H(X,Y)$ corresponds to their union.

uncertainties simply add up. In the general case of dependent variables we would have $H(X,Y) = H(X) + H(Y|X) = H(Y) + H(X|Y)$: the joint uncertainty is the sum of the uncertainty of one variable and the uncertainty of the other knowing the first.

At the other extreme, suppose Y is fully dependent on X so that $Y = f(X)$ where f is some deterministic function. Then $H(Y|X) = 0$ and the uncertainty Y set is contained inside the uncertainty X set in Figure 2. Therefore,

$$H\big(f(X)\big) \leq H(X). \tag{96}$$

This means that *processing reduces entropy*: any function of a random variable has the effect of decreasing its entropy.

Similarly, for any two random variables X and Y, both uncertainty sets in the Venn diagram become smaller for $f(X)$ and $g(Y)$ for any functions f and g. Therefore, we have

$$I\big(f(X);g(Y)\big) \leq I(X;Y). \tag{97}$$

In words, *data processing can only reduce information.* This is a particular instance of an important result in information theory known as the *data processing inequality* [9, Thm. 2.8.1] for Markov chains.

32. Shannon's Channel Coding Theorem

The mutual information, along with the asymptotic large deviation bound (66), can be used to solve the *information transmission* problem: How can we reliably transmit a source of information at the highest possible speed? This corresponds to the model of Figure 1 in which the information source is to be transmitted though the noisy channel at the maximum possible transmission rate while achieving an arbitrarily reliable communication.

A discrete sequence $\underline{x} = (x_1, x_2, \ldots, x_n)$ input to the channel normally corresponds to a chosen channel *code* and is not random. But Shannon had the brilliant idea to consider the whole set of *all possible codes* that may be used in the communication, and assign an (albeit artificial) probability to each code in such a way that $\underline{x}$ can be considered as a realization of an i.i.d. sequence, as if the code were chosen at random with independent code sequences. This is certainly the first application of the famous "probabilistic method" later attributed to Paul Erdős [16]. In this non-constructive method, known as *random coding*, Shannon considers the average performance over all "random" codes and deduces the existence of at least one "good" code. Roughly sketched, his argument is as follows.

Assume that the sequence $\underline{x} = (x_1, x_2, \ldots, x_n)$ is input to a memoryless noisy channel. Then the corresponding channel output $\underline{y} = (y_1, y_2, \ldots, y_n)$ has the property that it is jointly typical with the channel input $\underline{x}$ with high probability, in the sense of the law of large numbers (see § 13). Therefore, to achieve an arbitrarily reliable communication, it would be sufficient in theory to decode the received signal $\underline{y}$ by selecting the code sequence $\underline{x}$ that is jointly typical with it, provided that the actual transmitted code $\underline{x}$ is the only sequence having this property.

Now if another code sequence $\underline{x}'$ happens to be jointly typical with $\underline{y}$, since the code sequences are chosen to be independent, the actual probability distribution p of the corresponding bivariate random variable (X', Y) is $p(x)p(y)$ while its type q is roughly equal to the joint distribution $p(x, y)$. From § 24 and the asymptotic large deviation bound (66), the probability that this happens is bounded by $\exp\bigl(-nD(q,p)\bigr)$ where $D(q,p) = D(p(x,y), p(x)p(y)) = I(X;Y)$ by definition of mutual information (85). For a channel with N code sequences, the total decoding error probability P_e (averaged over all possible codes) is then bounded by

$$P_e \leq N \exp\bigl(-nI(X;Y)\bigr). \tag{98}$$

For this error probability to be exponentially small as $n \to +\infty$, it is sufficient that the transmission rate R per symbol be strictly less than the mutual information:

$$R = \frac{\log N}{n} < I(X,Y). \tag{99}$$

In order to maximize the transmission rate, the probability distribution $p(x)$ of code sequences can be chosen so as to maximize $I(X;Y)$. Shannon's channel *capacity*

$$C = \max_{p(x)} I(X;Y) \tag{100}$$

is the maximum possible amount of information transmitted over the communication channel. Thus, there exists a code achieving arbitrarily small probability of decoding error, provided that

$$R < C. \tag{101}$$

This is the celebrated *Shannon's second coding theorem* [43] which provides the best possible performance of any data transmission scheme over a noisy channel: An arbitrarily reliable communication can be achieved so long as the transmission rate does not exceed the channel's capacity.

This revolutionary theorem did change our world. For the first time, it was realized that the transmission noise does not limit the reliability of the communication, only the speed of transmission. Thus, digital communications can achieve almost perfect quality. That alone justifies that Shannon is considered as the father of the digital information age.

Somewhat paradoxically, even though Shannon's channel coding theorem is non-constructive, it suggest that any code picked at random would be very likely to be almost optimal. However, since $n \to +\infty$ in Shannon's argument, such a code would be impossible to implement in practice. Intensive research is still undertaken today to derive good channel codes, sufficiently complex (that appear 'random') to perform well and at the same time sufficiently simple to be efficiently implemented.

33. Shannon's Capacity Formula

Perhaps the most emblematic classical expression of information theory is Shannon's capacity formula for a communication channel with additive white Gaussian noise (AWGN). The AWGN model is the basic noise model used to mimic the effect of many random processes that occur in nature, and a very good model for many practical communication links. The capacity is given by (100) where in the AWGN model,

$$Y = X + Z^*, \tag{102}$$

where $Z^* \sim \mathcal{N}(0, N)$ is a Gaussian random variable with zero mean and power N, independent of the transmitted signal X. The maximum in (100) is to be taken over distributions $p(x)$ such that X has limited power P. Here the quantity $\mathrm{SNR} = P/N$ is known as the signal-to-noise ratio (SNR). We have

$$\begin{aligned} I(X;Y) &= h(Y) - h(Y|X) && \text{by (88)} && (103)\\ &= h(Y) - h(Z^*) && \text{by (102)} && (104)\\ &\leq h(Y^*) - h(Z^*) && \text{by (54),} && (105) \end{aligned}$$

where Y^* is a Gaussian random variable having the same power as $Y = X + Z^*$, that is, $P+N$. The upper bound (105) is attained when $X = X^*$ is itself Gaussian, since then $Y = X^* + Z^* = Y^*$ is also Gaussian (as the sum of independent Gaussian variables). Therefore, (105) is the required capacity. From (39) we obtain

$$C = \frac{1}{2}\log\bigl(2\pi e(P+N)\bigr) - \frac{1}{2}\log(2\pi e N) = \frac{1}{2}\log(1 + P/N). \tag{106}$$

This is the celebrated Shannon's capacity formula $C = (1/2)\log(1+\mathrm{SNR})$ that appears in Shannon's 1948 paper. It is often said that Hartley derived a similar rule twenty years before Shannon, but in fact this is a historical misstatement [38]. However, this formula was discovered independently by at least seven other researchers in the same year 1948! [38] An illustration of a concept whose time has come.

34. The Entropy Power Inequality and a Saddle Point Property

It is well known that the power of the sum of independent zero-mean random variables equals the sum of the individual powers. For the entropy power (42), however, with have the inequality

$$N(X+Y) \geq N(X) + N(Y) \tag{107}$$

for any independent variables X and Y. This is known as the *entropy power inequality* (EPI): The entropy power of the sum of independent random variables *is not less* than the sum of the individual entropy powers.

The EPI was first stated by Shannon in his 1948 paper and is perhaps the most difficult and fascinating inequality in the theory. Shannon's 1948 proof [43] was incomplete; the first rigorous proof was given ten years later by Stam [50], with a quite involved argument based on the Fisher information. In the last section of this paper I will present a simple and recent (2017) proof from [39].

The EPI was initially used by Shannon to evaluate the channel capacity for non-Gaussian channels. It now finds many applications in information theory (to bound performance regions for multi-user source or channel coding problems) and in mathematics (e.g., to prove strong versions of the central limit theorem).

It is particularly interesting to review how Shannon used the EPI to specify the role of the Gaussian distribution for communication problems. From the preceding section (§ 33), we know that the Gaussian X^* maximizes the mutual information $I(X;Y)$ transmitted over a channel with additive Gaussian noise Z^*, i.e.,

$$I(X;X+Z^*) \leq I(X^*;X^*+Z^*) = C. \tag{108}$$

where $C = \frac{1}{2}\log(1+P/N)$. On the other hand, when the additive noise Z of power N is not necessarily Gaussian and the channel input X^* is Gaussian of power P, we have

$$\begin{aligned} I(X^*;X^*+Z) &= h(X^*+Z) - h(Z) &(109)\\ &= \frac{1}{2}\log\bigl(2\pi e N(X^*+Z)\bigr) - \frac{1}{2}\log\bigl(2\pi e N(Z)\bigr), &(110) \end{aligned}$$

where by the EPI (107), $N(X^*+Z) \geq N(X^*) + N(Z) = P + N(Z)$. Therefore,

$$I(X^*;X^*+Z) \geq \frac{1}{2}\log\bigl(1+P/N(Z)\bigr) \geq \frac{1}{2}\log\bigl(1+P/N\bigr), \tag{111}$$

where we have used that the entropy power $N(Z)$ does not exceed the actual power N. Combining this with (108) we obtain a *saddle point property of mutual information*:

$$I(X;X+Z^*) \leq \underbrace{I(X^*;X^*+Z^*)}_{C=\frac{1}{2}\log(1+P/N)} \leq I(X^*;X^*+Z). \tag{112}$$

This shows that the *Gaussian* is at the same time the *best signal* X^* (which maximizes information) and the *worst noise* Z^* (which minimizes information).

From this result, one can define a two-person (signal X and noise Z) zero-sum game with mutual information as the payoff function for which a *Nash equilibrium* holds with the Gaussian saddle point (X^*, Z^*) [1]. Similar considerations can be used to establish a certain duality between source and channel coding.

35. MaxEnt vs. MinEnt Principles

Before going through the proof the EPI (107), it is convenient to rewrite it as an extremum property of the differential entropy. Let X and Y be any two independent continuous random variables and let X^* and Y^* be zero-mean Gaussian independent variables.

From the *maximum entropy (MaxEnt) principle* (see (54) in § 21) we know that under a fixed variance constraint, the differential entropy is maximized for a Gaussian variable. Thus under the condition of identical individual *variances*:

$$\mathsf{Var}(X^*) = \mathsf{Var}(X) \quad \text{and} \quad \mathsf{Var}(Y^*) = \mathsf{Var}(Y), \tag{113}$$

the entropy of the linear combination $aX + bY$ is *maximized* for Gaussian variables:

$$h(aX + bY) \leq h(aX^* + bY^*). \tag{114}$$

This is simply a consequence of the fact that $aX^* + bY^*$ is Gaussian (as the sum of independent Gaussian variables) of the same variance as $aX + bY$.

Interestingly, the EPI (107) can be rewritten as a *minimum entropy (MinEnt) principle*: Under the condition of identical individual *entropies*:

$$h(X^*) = h(X) \quad \text{and} \quad h(Y^*) = h(Y), \tag{115}$$

the entropy of the linear combination $aX + bY$ is now *minimized* for Gaussian variables:

$$h(aX + bY) \geq h(aX^* + bY^*). \tag{116}$$

To see this, notice that from the scaling property of the entropy power (43), the EPI (107) can be rewritten as $N(aX+bY) \geq N(aX)+N(bY) = a^2N(X)+b^2N(Y)$. But from (115) we have $N(X) = N(X^*)$ and $N(Y) = N(Y^*)$. Since the entropy power of a Gaussian variable is the same as its power, $a^2N(X) + b^2N(Y) = a^2N(X^*) + b^2N(Y^*) = N(aX^* + bY^*)$. Thus the EPI can be rewritten as $N(aX + bY) \geq N(aX^* + bY^*)$ which taking logarithms is the same as (116).

In addition, we shall see that the minimum is achieved only for Gaussian variables provided the linear combination is not trivial (a, b are non-zero scalars). This MinEnt form (116) of the EPI finds application in signal processing for blind source separation and deconvolution.

36. A Simple Proof of the Entropy Power Inequality [39]

Lastly, proceed to prove the EPI in the form (116). First, by the scaling property of entropy (27), we can always modify the constants a, b in such a way that X and

Y can be assumed to have equal entropies. Then condition (115) writes

$$h(X) = h(Y) = h(X^*) = h(Y^*). \tag{117}$$

In particular, the zero-mean Gaussian variables X^* and Y^* have equal entropies, equal variances, and, therefore, identical normal distributions.

Next, applying by the scaling property of entropy (27) if necessary in (116), we can always assume that a, b have been further normalized such that $a^2 + b^2 = 1$. Then $\widetilde{X} = aX^* + bY^*$ in the right-hand side of (116) is also identically distributed as X^* and Y^*. In fact, the rotation

$$\begin{pmatrix} \widetilde{X} \\ \widetilde{Y} \end{pmatrix} = \begin{pmatrix} a & b \\ -b & a \end{pmatrix} \begin{pmatrix} X^* \\ Y^* \end{pmatrix} \tag{118}$$

transform i.i.d. Gaussian variables X^*, Y^* into i.i.d. Gaussian variables $\widetilde{X}, \widetilde{Y}$.

We now use the classical *inverse transform sampling* method: Let Φ be the cumulative distribution function (c.d.f.) of X^* and F be the c.d.f. of X. Then

$$\begin{aligned} F(x) &= \Phi\left(\Phi^{-1}(F(x))\right) &(119)\\ &= \mathbb{P}\left(X^* \leq \Phi^{-1}(F(x))\right) &(120)\\ &= \mathbb{P}\big(\Phi(X^*) \leq F(x)\big) &(121)\\ &= \mathbb{P}\left(F^{-1}(\Phi(X^*)) \leq x\right). &(122) \end{aligned}$$

Thus letting $T = F^{-1} \circ \Phi$, the variable $T(X^*)$ has the same distribution as X, and we can write $X = T(X^*)$ in the above expressions. This so-called *transport argument* can also be applied to Y. Thus, we can always assume that

$$X = T(X^*) \quad \text{and} \quad Y = U(Y^*) \tag{123}$$

for some "transportation" functions T and U. The EPI (116) now writes

$$h\big(aT(X^*) + bU(Y^*)\big) \geq h(aX^* + bY^*), \tag{124}$$

where the right-hand side equals $h(\widetilde{X})$. By the inverse of rotation (118)

$$\begin{pmatrix} X^* \\ Y^* \end{pmatrix} = \begin{pmatrix} a & -b \\ b & a \end{pmatrix} \begin{pmatrix} \widetilde{X} \\ \widetilde{Y} \end{pmatrix}, \tag{125}$$

the EPI (116) is equivalent to the inequality:

$$h\big(a\,T(a\widetilde{X} - b\widetilde{Y}) + b\,U(b\widetilde{X} + a\widetilde{Y})\big) \geq h(\widetilde{X}). \tag{126}$$

We now proceed to prove (126). Since conditioning reduces entropy (93),

$$h\big(a\,T(a\widetilde{X} - b\widetilde{Y}) + b\,U(b\widetilde{X} + a\widetilde{Y})\big) \geq h\big(a\,T(a\widetilde{X} - b\widetilde{Y}) + b\,U(b\widetilde{X} + a\widetilde{Y})\big|\widetilde{Y}\big). \tag{127}$$

Apply the change of variable formula (25) in the right-hand side, where $\widetilde{Y}$ is fixed so that $T_{\widetilde{Y}}(\widetilde{X}) = a\,T(a\widetilde{X} - b\widetilde{Y}) + b\,U(b\widetilde{X} + a\widetilde{Y})$ is a function of $\widetilde{X}$ alone with derivative $a^2T'(a\widetilde{X}-b\widetilde{Y})+b^2U'(b\widetilde{X}+a\widetilde{Y})$. Then

$$\begin{aligned} h\big(a\,T(a\widetilde{X}-b\widetilde{Y})+b\,U(b\widetilde{X}+a\widetilde{Y})\big|\widetilde{Y}\big) \\ &= h(\widetilde{X}\mid\widetilde{Y})+\mathbb{E}\log\big(a^2T'(a\widetilde{X}-b\widetilde{Y})+b^2U'(b\widetilde{X}+a\widetilde{Y})\big) && (128)\\ &= h(\widetilde{X})+\mathbb{E}\log\big(a^2T'(X^*)+b^2U'(Y^*)\big). && (129)\end{aligned}$$

It remains to prove that the second term $\mathbb{E}\log\big(a^2T'(X^*)+b^2U'(Y^*)\big)$ is nonnegative. By the concavity property of the logarithm,

$$\begin{aligned}\mathbb{E}\log\big(a^2T'(X^*)+b^2U'(Y^*)\big) &\geq \mathbb{E}\big(a^2\log T'(X^*)+b^2\log U'(Y^*)\big) && (130)\\ &= a^2\,\mathbb{E}\log T'(X^*)+b^2\,\mathbb{E}\log U'(Y^*), && (131)\end{aligned}$$

where from (25) and (117),

$$\begin{aligned}a^2\,\mathbb{E}\log T'(X^*)+b^2\,\mathbb{E}\log U'(Y^*) &= a^2\big(h(T(X^*))-h(X^*)\big)+b^2\big(h(U(Y^*))-h(Y^*)\big)\\ &= a^2\big(h(X)-h(X^*)\big)+b^2\big(h(Y)-h(Y^*)\big) && (132)\\ &= 0. && (133)\end{aligned}$$

This ends the proof of the EPI.

The equality case in (116) can be easily settled in this proof. If the linear combination is not trivial (that is, if both a and b are nonzero scalars), equality holds in the concavity inequality (130) if and only if $T'(X^*)=U'(Y^*)$. Since X^* and Y^* are independent Gaussian variables, this implies that the derivatives T' and U' are constant and equal, hence X and Y in (123) are Gaussian. Thus equality holds in (116) only for Gaussian variables.

37. Conclusion

Who else but Shannon himself can conclude on his life's work in information theory?

> *"I didn't think in the first stages that it was going to have a great deal of impact. I enjoyed working on this kind of a problem, as I have enjoyed working on many other problems, without any notion of either financial gain or in the sense of being famous; and I think indeed that most scientists are oriented that way, that they are working because they like the game."* [49]

References

[1] N.M. Blachman, "Communication as a Game," 1957 IRE WESCON Convention Record, pt. 2, 1957, pp. 61–66.

[2] Ludwig Boltzmann, "Weitere Studien über das Wärmegleichgewicht unter Gasmolekülen," *Sitzungsberichte Kaiserlichen Akademie der Wissenschaften*, Wien Mathematisch Naturwissenschaftliche Classe, Vol. 66, pp. 275–370, 1872.

[3] Léon Brillouin, *Science and Information Theory*, Academic Press: New York, U.S.A., 1956.

[4] Samuel Hawksley Burbury, "Boltzmann's Minimum Function," *Nature*, Vol. 51, No. 1308, Nov. 1894, p. 78.

[5] John Parker Burg, *Maximum Entropy Spectral Analysis*, Ph.D. thesis, Department of Geophysics, Stanford University, Stanford, California, U.S.A., May 1975.

[6] Herman Chernoff, "A Measure of the Asymptotic Efficiency of Tests of a Hypothesis Based on a Sum of Observations," *Annals of Mathematical Statistics*, Vol. 23, No. 4, 1952, pp. 493–507.

[7] Adrian Cho, "A Fresh Take on Disorder, Or Disorderly Science?," *Science*, Vol. 297, No. 5585, 2002, pp. 1268–1269.

[8] Rudolf Clausius, *The Mechanical Theory of Heat – with its Applications to the Steam Engine and to Physical Properties of Bodies*, Ninth Memoir (1865): "On Several Convenient Forms of the Fundamental Equations of the Mechanical Theory of Heat," pp. 327–365, 1865.

[9] Thomas M. Cover and Joy A. Thomas, *Elements of Information Theory*, John Wiley & Sons, 2nd ed., 2006.

[10] Harald Cramér, *Mathematical Methods of Statistics*, Princeton University Press.: Princeton, NJ, U.S.A., 1946.

[11] Imre Csiszár, "Eine informationstheoretische Ungleichung und ihre Anwendung auf den Beweis der Ergodizität von Markoffschen Ketten". Magyar. Tud. Akad. Mat. Kutato Int. Kozl., Vol. 8, 1963, pp. 85–108.

[12] Georges Darmois, "Sur les limites de la dispersion de certaines estimations," *Revue de l'Institut International de Statistique*, Vol. 13, 1945, pp. 9–15.

[13] Joseph Leo Doob, "Review of C. E. Shannon's 'A Mathematical Theory of Communication'," *Mathematical Reviews*, vol. 10, p. 133, Feb. 1949.

[14] John P. Dougherty, "Foundations of Non-equilibrium Statistical Mechanics," *Philosophical Transactions: Physical Sciences and Engineering*, Royal Society London, Vol. 346, No. 1680, pp. 259–305, Feb. 1994.

[15] F.Y. Edgeworth, "On the Probable Errors of Frequency-Constants," *Journal of the Royal Statistical Society*, Vol. 71, No. 2,3,4, 1908, pp. 381–397, 499–512, 651–678.

[16] Paul Erdős, "Graph Theory and Probability," *Canadian Journal of Mathematics*, Vol. 11, No. 0, 1959, pp. 34–38.

[17] Robert Mario Fano, *Transmission of Information: A Statistical Theory of Communications*. Cambridge, Mass: MIT Press, 1961.

[18] ———, interview by Arthur L. Norberg, Charles Babbage Institute, Center for the History of Information Processing, University of Minnesota, Minneapolis, 20–21 April 1989. See also [17, p. vii].

[19] ———, interview by Aftab, Cheung, Kim, Thakkar, Yeddanapudi, 6.933 Project History, Massachusetts Institute of Technology, Nov. 2001.

[20] Ronald Aylmer Fisher, "On the Mathematical Foundations of Theoretical Statistics," *Philosophical Transactions of the Royal Society of London A* Vol. 222, 1922, pp. 309–368.

[21] Maurice Fréchet, “Sur l'extension de certaines évaluations statistiques au cas de petit échantillons,” *Revue de l'Institut International de Statistique*, Vol. 11, 1943, pp. 182–205.

[22] Howard Gardner, *The Mind's New Science: A History of the Cognitive Revolution.* Basic Books, 1987, p. 144.

[23] Josiah Willard Gibbs, *Elementary Principles in Statistical Mechanics (developed with especial reference to the rational foundation of thermodynamics)*, Dover Publications: New York, U.S.A., 1902. (Theorem I p. 129)

[24] Friedrich-Wilhelm Hagemeyer, *Die Entstehung von Informationskonzepten in der Nachrichtentechnik: Eine Fallstudie zur Theoriebildung in der Technik in Industrie und Kriegsforschung* [The Origin of Information Theory Concepts in Communication Technology: Case Study for Engineering Theory-Building in Industrial and Military Research], Doctoral Dissertation, Freie Universität Berlin, Nov. 8, 1979, 570 pp.

[25] Ralph Vinton Lyon Hartley, “Transmission of Information,” *Bell System Technical Journal*, July 1928, pp. 535–563.

[26] Jan Havrda and František Charvát, “Quantification Method of Classification Processes: Concept of Structural α-Entropy,” *Kybernetika*, Vol. 3, No. 1, 1967, pp. 30–34.

[27] Erik Hollnagel and David D. Woods, *Joint Cognitive Systems: Foundations of Cognitive Systems Engineering*, Taylor & Francis: Boca Raton, FL, U.S.A., 2005 (p. 11).

[28] Stig Hjalmars, “Evidence for Boltzmann's H as a Capital Eta”, *American Journal of Physics*, Vol. 45, No. 2, Feb. 1977, p. 214.

[29] Edwin Thompson Jaynes, “Information Theory and Statistical Mechanics,” *Physical Review* Vol. 106, No. 4, pp. 620–630, May 1957 and Vol. 108, No. 2, pp. 171–190, Oct. 1957.

[30] Yu.L. Klimontovich, *Statistical Theory of Open Systems. I. A Unified Approach to Kinetic Description of Processes in Active Systems.* Kluwer Academic Publishers: Dordrecht, The Netherlands, 1995 (p. 25).

[31] Ronald R. Kline, *The Cybernetics Moment: Or Why We Call Our Age the Information Age.* Johns Hopkins University Press: Baltimore, MD, U.S.A., 2015, xi+ 336 pp. (p. 123).

[32] Andreï Nikolaïevitch Kolmogorov, “Combinatorial Foundations of Information Theory and the Calculus of Probabilities,” talk at the *International Mathematical Congress* (Nice, 1970), *Russian Mathematical Surveys*, Vol. 38, No. 4, pp. 29–40, 1983.

[33] S Kullback and R.A. Leibler, “On Information and Sufficiency,” *Annals of Mathematical Statistics*, Vol. 22, No. 1, pp. 79–86, 1951.

[34] Harry Nyquist, "Certain Factors Affecting Telegraph Speed,” *Bell System Technical Journal*, April 1924, pp. 324–346, and “Certain Topics in Telegraph Transmission Theory,” *Transactions of the American Institute of Electrical Engineers*, Vol. 47, April 1928, pp. 617–644.

[35] John Robinson Pierce, “The Early Days of Information Theory,” *IEEE Transactions on Information Theory*, vol. IT-19, no. 1, Jan. 1973, pp. 3–8.

[36] Calyampudi Radakrishna Rao, “Information and the Accuracy Attainable in the Estimation of Statistical Parameters,” *Bulletin of the Calcutta Mathematical Society*, Vol. 37, 1945, pp. 81–89.

[37] Alfréd Rényi, "On Measures of Entropy and Information," in *Proc. 4th Berkeley Symp. Math., Stat. Prob.*, Berkeley, California, U.S.A., 20 June–30 July 1960, University of California Press, Vol. 1, 1960, pp. 547–561.

[38] Olivier Rioul and José Carlos Magossi, "On Shannon's Formula and Hartley's Rule: Beyond the Mathematical Coincidence," in *Entropy*, Special Issue on *Information, Entropy and their Geometric Structures*, Vol. 16, No. 9, pp. 4892–4910, Sept. 2014.

[39] Olivier Rioul, "Yet Another Proof of the Entropy Power Inequality," *IEEE Transactions on Information Theory*, Vol. 63, No. 6, June 2017, pp. 3595–3599.

[40] Claude Elwood Shannon, "A Symbolic Analysis of Relay and Switching Circuits," Thesis (Master of Science), M.I.T., August 10, 1937. In *Transactions American Institute of Electrical Engineers*, Vol. 57, 1938, pp. 713–723.

[41] ——, "An Algebra for Theoretical Genetics," Ph.D. Dissertation, Department of Mathematics, Massachusetts Institute of Technology, April 15, 1940, 69 pp.

[42] ——, "A Mathematical Theory of Cryptography," Memorandum MM 45-110-02, Sept. 1, 1945, Bell Laboratories, 114 pp. Declassified and reprinted as "Communication Theory of Secrecy Systems," *Bell System Technical Journal*, Vol. 28, 1949, pp. 656–715.

[43] ——, "A Mathematical Theory of Communication," *Bell System Technical Journal*, Vol. 27, July and October 1948, pp. 379–423 and 623–656.

[44] —— (with Warren Weaver), *The Mathematical Theory of Communication*, University of Illinois Press, Urbana, IL, U.S.A., 1949, vi+117 pp.

[45] ——, "The Bandwagon," *IRE Transactions on Information Theory*, Editorial, Vol. 2, No. 1, March 1956, p. 3.

[46] ——, interview by Friedrich-Wilhelm Hagemeyer, Winchester, Massachusetts, U.S.A., February 28, 1977.

[47] ——, interview by Robert Price, Winchester, Massachusetts, U.S.A., IEEE History Center, July 28, 1982. Partly published in F. W. Ellersick, "A Conversation with Claude Shannon," IEEE Communications Magazine, Vol. 22, 1984, pp. 123–126.

[48] ——, interview by Anthony Liversidge, *Omni magazine*, August 1987.

[49] ——, TV interview, circa 1990.

[50] Adriaan Johannes Stam, "Some Inequalities Satisfied by the Quantities of Information of Fisher and Shannon," *Information and Control*, Vol. 2, No. 2, 1959, pp. 101–112.

[51] Leó Szilárd, "Über die Entropieverminderung in einem thermodynamischen System bei Eingriffen intelligenter Wesen." [On the Decrease of Entropy in a Thermodynamic System by the Intervention of Intelligent Beings], *Zeitchrift für Physik*, Vol. 53, 1929, 840–856.

[52] Libb Thims, "Thermodynamics ≠ Information Theory: Science's Greatest Sokal Affair," *Journal of Human Thermodynamics*, Vol. 8, No. 1, Dec. 2012, pp. 1–120.

[53] Myron Tribus and Edward C. McIrvine, "Energy and Information," *Scientific American*, Vol. 225, 1971, pp. 179–188.

[54] Myron Tribus, "A Tribute to Edwin T. Jaynes," in *Proceedings of the 18th International Workshop on Maximum Entropy and Bayesian Methods of Statistical Analysis*, Garching, Germany, 1998, pp. 11–20.

[55] Constantino Tsallis, "Possible Generalization of Boltzmann-Gibbs Statistics," *Journal of Statistical Physics*, Vol. 52, Nos. 1/2, 1988, pp. 479–487.

[56] Johann von Neumann, *Mathematische Grundlagen der Quantenmechanik*, Verlag Von Julius Springer: Berlin, Germany, 1932.

[57] Warren Weaver, "The Mathematics of Communication," *Scientific American*, vol. 181, no. 1, July 1949, pp. 11–15.

[58] Norbert Wiener, "The Extrapolation, Interpolation, and Smoothing of Stationary Time Series (with Engineering Applications)," M.I.T., Feb. 1, 1942. Published by Technology Press and John Wiley & Sons, 1949.

[59] ———, *Cybernetics*, Chapter III: Time series, Information and Communication, John Wiley & Sons: New York, NY, U.S.A., 1948, pp. 10–11.

[60] ———, "What is Information Theory?" *IRE Transactions on Information Theory*, Editorial, Vol. 2, No. 2, July 1956, p. 48.

[61] IEC 80000-13:2008, Quantities and units – Part 13: Information science and technology, International Organization for Standardization.

Olivier Rioul
LTCI, Télécom ParisTech
Université Paris-Saclay
75013 Paris, France
and
CMAP
École Polytechnique
Université Paris-Saclay
91128 Palaiseau, France
e-mail: olivier.rioul@telecom-paris.fr

Information Theory, 87–112

Poincaré Seminar 2018

Landauer's Bound and Maxwell's Demon

Sergio Ciliberto

Abstract. We summarize recent experimental and theoretical progress that has been achieved in the physics of information. We highlight the intimate connection existing between information and energy from Maxwell's demon and Szilard's engine to Landauer's erasure principle. We will focus mainly on experiments on classical systems and we will shortly discuss a few aspects of quantum systems. We conclude by discussing applications in engineering and biology.

1. Introduction

This review summarizes the contents of several articles [1, 2, 3, 4, 5], that we wrote on the experimental aspects of the connections between thermodynamics and information. We first define the theoretical framework of this connection by presenting a short historical review starting from the article [6] of Rolf Landauer in which he argued that information is physical. Since information is processed in physical devices, he concluded that information has to obey the laws of physics, and in particular the laws of thermodynamics. Information is thus stored in physical systems, such as books or memory sticks, and transmitted by physical means, for instance with the help of electrical or optical signals. But what is 'information'? A simple, intuitive answer is 'what you don't already know'. If someone tells you that the earth is spherical, you surely would not learn much: this message has low information content. However, if you are told that the oil price will double the day after tomorrow, assuming for a moment this to be true, you would learn a great deal: this message has hence high information content. Mathematically, the amount of information is quantified by the so-called information entropy H introduced by Claude Shannon in 1948; the larger the entropy, the bigger the information content [7]. The simplest device to store information is a system with two distinct states, for example up/down, left/right or magnetization/no magnetization. If the system is known to be with probability one in one of either states, probing the system will not reveal any new information, and the Shannon entropy is zero. On the other hand, if the two states can be occupied with probability one-half, and the actual state is therefore initially undetermined, an examination of the system will provide information about the state it is in. In this case, the Shannon entropy is equal to ln(2). This value corresponds to the smallest amount of information and is called a bit. A two-state system can thus store up to one bit of information.

The second law of thermodynamics, as formulated by Rudolf Clausius in 1850, is based on the empirical observation that some processes only occur spontaneously in one preferred direction [8]. Everyone who forgot a cup of hot tea on a table has noted that heat flows by itself from a hotter (the cup) to a colder body (the room), and never the other way around. Heat flow is therefore said to be irreversible. Clausius characterized the irreversibility of natural macroscopic processes by defining the thermodynamic entropy S, a quantity that is not conserved, in contrast to energy, but can only increase in isolated systems. This asymmetry in the change of entropy imposes restrictions on the type of physical phenomena that are possible. Similarly, the application of the second law of thermodynamics to information sets limitations on information processing tasks such as transmission or erasure. More general questions address the thermodynamic consequences of information gain. In particular, whether it is possible to extract useful mechanical work from a system by observing its state, and if yes how much. And at the more fundamental level: are thermodynamic and information entropies related [2, 9]?

1.1. Maxwell's Demon and Szilard's Engine

The first hint of a connection between information and thermodynamics may be traced back to James Clerk Maxwell's now famous demon introduced in 1867 [10, 11, 12]. The demon is an intelligent creature able to monitor individual molecules of a gas contained in two neighboring chambers initially at the same temperature, as shown in Fig. 1. The temperature of the gas is defined by the mean kinetic energy of the molecules and is hence proportional to their mean-square velocity. However, not all the particles will have the same velocity. Some of the molecules will be going faster than average and some will be going slower. By opening and closing a molecular-sized trap door in the partitioning wall, the demon collects the faster molecules in one of the chambers and the slower ones in the other. The two chambers now contain gases with different mean-square velocities and hence different temperatures. This temperature difference may be used to run a heat engine and produce mechanical work. By gathering information about the position and velocity of each particle and using this knowledge to sort them, the demon is therefore able to decrease the entropy of the system and convert the acquired information into energy. The problem is that the demon, assuming a frictionless trap door, is able to do all this without performing any work himself, in apparent violation of the second law of thermodynamics. The proper resolution of this paradox took 115 years.

A simplified one-particle engine has been suggested by Leo Szilard in 1929 [13]. In this setup, schematically shown in Fig. 2, the gas consists of a single molecule and the wall separating the identical chambers is replaced by a moving

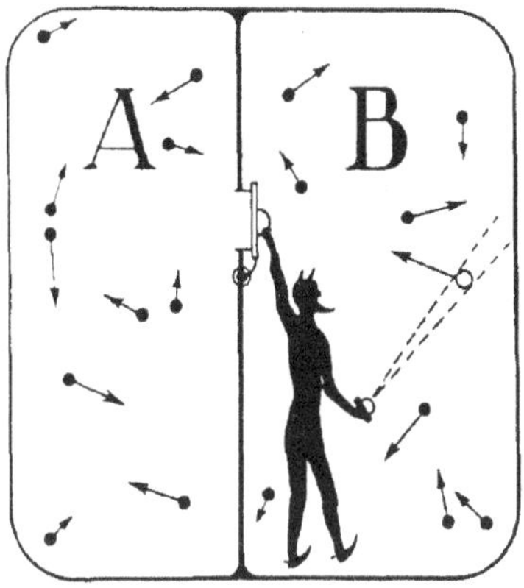

FIGURE 1. Maxwell's demon. By detecting the positions and velocities of gas molecules in two neighboring chambers and using that information to time the opening and closing of a trapdoor that separates them, a tiny, intelligent being could, in theory, sort molecules by velocity. By doing so, it could create a temperature difference across the chambers that could be used to perform mechanical work. If the trapdoor is frictionless, the sorting requires no work from the demon himself, in apparent violation of the second law of thermodynamics (drawn by Claire Lebeau)

piston to which a weight can be attached. We now have a two-state system very similar to the one discussed above. Initially, the particle has a probability of one half to be in one of the two chambers. By looking into the container the demon acquires information about the actual state of the system, learning what he did not know before. If the molecule is found in the right chamber, the weight is attached to the right-hand side of the piston which is then released from its former position. During the expansion of the gas, the piston is pushed to the left and the weight is pulled upwards, performing work against gravity. The weight is attached to the left-hand side of the piston when the molecule is observed in the left chamber. The second law of thermodynamics limits the maximum amount of work that can be produced by the Szilard engine to $k_B T \ln(2)$, where k_B is the Boltzmann constant and T the temperature of the gas. This corresponds to the maximum amount of energy that can be obtained by converting one bit of information, and is historically the first clear statement of the relationship between information and energy. In modern language, this result further implies that information and thermodynamic

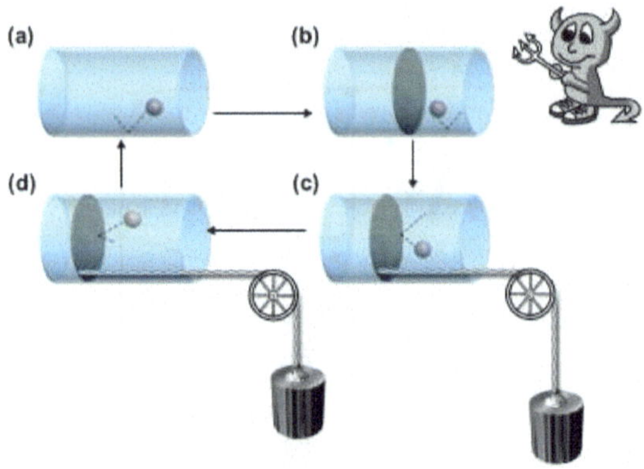

FIGURE 2. Szilard's engine. A crafty observer can turn a single particle in a box into an engine that converts information into mechanical work. If, say, the particle is found on the box's left-hand side, the observer inserts a movable wall and attaches a weight to its left side. The free expansion of the one-particle gas pushes the wall to the right, lifts the weight, and thereby performs work against gravity (adapted from Ref. [12]).

entropies are equal, $S = k_B H$, up to the multiplicative factor k_B introduced for dimensional reasons (the Shannon entropy H is dimensionless).

1.2. Landauer's Principle and Bennett's Resolution

It is useful to distinguish two complementary aspects: the first one is information gain, as we have just discussed with Maxwell's demon, the second one is information erasure, which has been investigated from a thermodynamic point of view by Landauer in 1961 (see Box 1). Let us again consider a two-state system and let us assume that it initially stores one bit of information, that is, the two states are occupied with equal probability one-half. This bit may be erased by resetting the system to one of the states, which will then be occupied with unit probability, a situation that corresponds to a zero Shannon entropy. By applying the second law of thermodynamics, Landauer demonstrated that information erasure is necessarily a dissipative process: the erasure of one bit of information is accompanied by the production of at least $k_B T \ln(2)$ of heat into the environment. This result is known as Landauer's erasure principle. It emphasizes the fundamental difference between the process of writing and erasing information. Writing is akin to copying information from one device to another: state left is mapped to left and state right is mapped to right, for example. This one-to-one mapping can be realized in principle without dissipating any heat (in statistical mechanics one would say that it conserves the volume in phase space). By contrast, erasing information is a two-to-one transformation: states left and right are mapped onto one single state, say right (this process does not conserve the volume in phase space and is thus dissipative).

Landauer's principle played a central role in solving the paradox of Maxwell's demon. In 1982 Charles Bennett noted that the demon has to store the information he acquires about the gas molecules in a memory [14]. After a full information gathering energy producing cycle, this memory has to be reset to its initial state to allow for a new iteration, and its information content has thus to be erased (a similar argument was put forward by Oliver Penrose in 1970 [15]). According to Landauer's principle, the erasure process will dissipate an amount of energy that is always larger than the quantity of energy produced by the demon during one cycle. The demon has consequently to pay an energetic price to sort the molecules and have heat flow from the colder chamber to the hotter chamber, in full agreement with the second law of thermodynamics. Before Bennett's resolution, it was often believed, following arguments put forward by Leon Brillouin and Dennis Gabor, that it was the energetic price of the measurement, that is, of the act of gathering information, that would save the second law [16]. However, as shown by Bennett, there is no fundamental energetic limitation on the measurement process, which like the copy operation, may in principle be performed without dissipation, in stark contrast to erasure.

Box 1: Landauer's erasure principle

Landauer's principle can be seen as a direct consequence of the second law of thermodynamics. Consider a system (SYS) coupled to a reservoir (RES) at temperature T. According to the second law, the total entropy change for system and reservoir is positive: $S_{\text{TOT}} = S_{\text{SYS}} + S_{\text{RES}} \geq 0$. Since the reservoir is always at equilibrium, owing to its very large size, we have following Clausius, $\Delta S_{\text{RES}} = Q_{\text{RES}}/T$. In other words, the heat absorbed by the reservoir satisfies $Q_{\text{RES}} \geq T\Delta S_{\text{SYS}}$. For a two-state system that stores one bit of information, there are initially two possible states that can be occupied with probability one half, and the initial Shannon entropy is $H_i = \ln(2)$. After erasure, the system is with unit probability in one of the states and the final Shannon entropy vanishes $H_f = 0$. The change of information entropy is thus $\Delta H = -\ln(2)$. During this erasure process the ability of the system to store information has been modified. By further using the (assumed) equivalence between thermodynamic entropy S and information entropy H we can write $\Delta S_{\text{SYS}} = k_B H = k_B \ln(2)$. We hence obtain $Q_{\text{RES}} \geq k_B T \ln(2)$, showing that the heat dissipated into the reservoir during the erasure of one bit of information is always larger than $k_B T \ln(2)$.

2. Experimental Implementations

For almost a century and a half, the demon belonged to the realm of a gedanken experiment as the tracking and manipulation of individual microscopic particles was impossible. However, owing to the remarkable progress achieved in the last decades, such experiments have now become feasible. Just to give a hint on what can be done, we will discuss in the following sections several experimental realizations of Maxwell's demon and Szilard's engine, as well as several verification of Landauer's principle.

2.1. Experiments on Maxwell's Demon

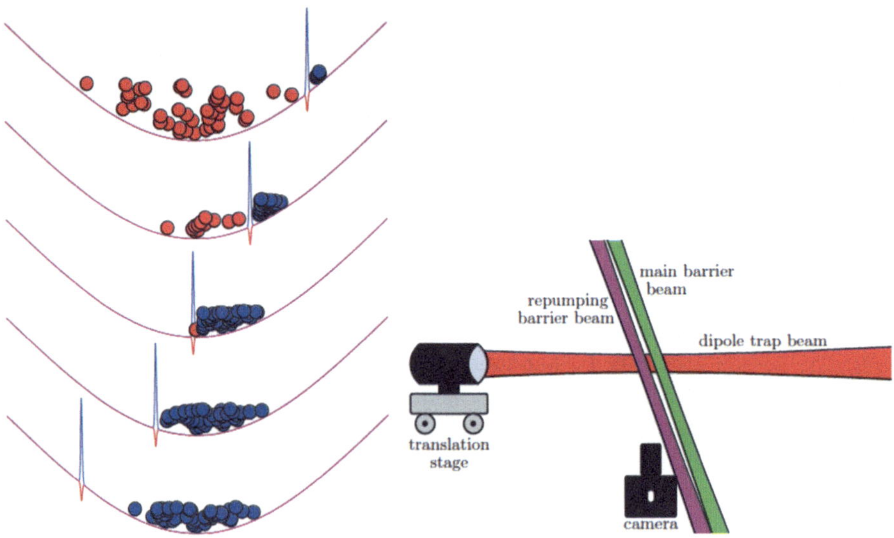

FIGURE 3. Using a Maxwell's demon to cool atoms. A pair of laser beams can be tuned to atomic transitions and configured to create a one-way potential barrier; atoms may cross unimpeded in one direction, from left to right left in this figure, but not in the other. Left panel : when the barrier is introduced at the periphery of the trapping potential, (right side) the atoms that cross the barrier will be those that have converted nearly all their kinetic energy to potential energy, in other words, the cold ones. By slowly sweeping the barrier (from the right to the left) across the trapping potential, one can sort cold atoms (blue) from hot ones (red), reminiscent of Maxwell's famous thought experiment, or cool an entire atomic ensemble. Because the cold atoms do work against the optical barrier as it moves, their kinetic energy remains small even as they return to the deep portion of the potential well. Right panel: schematic representation of the optical set-up showing the optical trap (red beam), the translational stage and the two beams one way barrier (adapted from Ref. [17]).

The first realization of a Maxwell demon was used to cool atoms in a magnetic trap. An ensemble of atoms is first trapped in a magnetic trap (see Fig. 3) [18]. A one way barrier (which plays the role of the demon) sweeps the magnetic trap from the right to the left, starting at a very large value of the potential. The atoms reaching this position have transformed almost all their kinetic energy in potential energy and are, therefore, very cool. These atoms go through the barrier but they cannot come back, i.e., the barrier behaves as an atom-diode [18, 19, 17]. Thus the hot atoms are on the right and the cold atoms are on the left. At the end of the process when the sweeping one-way barrier reaches the bottom of the magnetic potential all of the atoms are cooled down. The one way barrier is composed by two laser beams suitable tuned to atomic transitions. With reference to fig. one of the two lasers is on the left of the barrier and forces the atoms in an excited state. The frequency of the second laser, which is on the right of the barrier, is tuned in such a way that it has no effect on the atoms in the excited state and it repels the atoms in the ground state. Thus the atoms coming from the right, which are prepared in the excited state, go through the barrier and relax to the ground state by emitting a photon. Instead the atoms coming from the left, which are in the ground state, encounter first the barrier and remain trapped because they are repelled. Where does the connection with Maxwell demon come from? Indeed each time that an atom looses a photon the entropy of the light shining the atoms increases because before all the photons were coherently in the laser beam (low entropy state) and now the emitted photons are scattered in all directions (high entropy state). This entropy is related to an information entropy because each time that a photon is emitted we know that an atom has been cooled. It can be shown that indeed this gain of entropy is larger than the reduction of entropy produced by the cooling of the atomic cloud. It is important to notice that in this example the demon has not to be an intelligent being but it is just a suitable tuned device which automatically implements the operation.

2.1.1. *The Szilard engine: work production from information.* A Szilard engine has been realized in 2010 by using a single microscopic Brownian particle in a fluid and confined to a spiral-staircase-like potential shown in Fig. 4 [20]. Driven by thermal fluctuations, the particle performs an erratic up and down motion along the staircase. However, because of the potential gradient downwards steps will be more frequent than upwards steps and the particle will on average fall down. The position of the particle is measured with the help of a CCD camera. Each time the particle is observed to jump upwards, this information is used to insert a potential barrier that hinders the particle to move down. By repeating this procedure, the average particle motion is now upstairs and work is done against the potential gradient. By lifting the particle mechanical work has therefore been produced by gathering information about its position. This is the first example of a device that converts information into energy for a system coupled to a single thermal environment. However there is not a contradiction with the second law because Sagawa and Ueda formalized the idea that information gained through microlevel

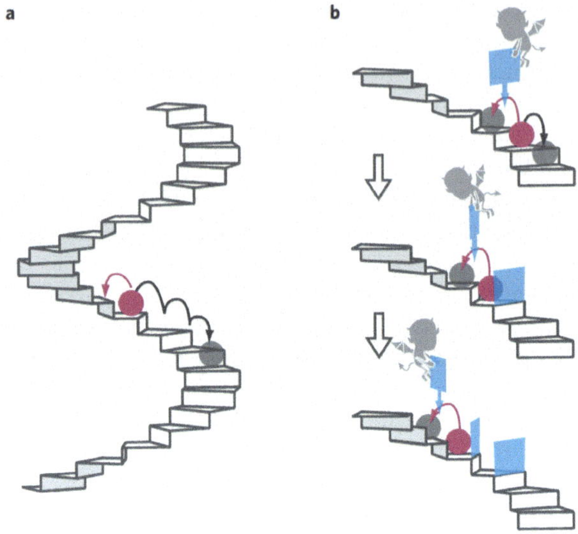

FIGURE 4. (a) Experimental realization of Szilard's engine. (a) A colloidal particle in a staircase potential moves downwards on average, but energy fluctuations can push it upwards from time to time. (b) When the demon observes such an event, he inserts a wall to prevent downward steps. By repeating this procedure, the particle can be brought to move upwards, performing work against the force created by the staircase potential. In the actual experiment, the staircase potential is implemented by a tilted periodic potential and the insertion of the wall is simply realized by switching the potential, replacing a minimum (no wall) by a maximum (wall) (adapted from Ref. [20]).

measurements can be used to extract added work from a heat engine. [21] Their formula for the the maximum extractable work is:

$$\langle W_{max} \rangle = -\Delta F + k_B\, T\, \langle I \rangle, \tag{1}$$

where ΔF is the free energy difference between the final and initial state and the extra term represents the so-called mutual information I. In absence of measurement errors this quantity reduces to the Shannon entropy : $I = -\sum_k P(\Gamma_k) \ln[P(\Gamma_k)]$, where $P(\Gamma_m)$ is the probability of finding the system in the state Γ_k. Then in the specific case of the previously described staircase potential [20]: $I = -p \ln p - (1-p) \ln p$ where p is the probability of finding the particle in a specific region.

In this context the Jarzynski equality (see appendix A) also contains this extra term and it becomes :

$$\langle \exp(-\beta W + I)\rangle \ = \exp(-\beta \Delta F), \tag{2}$$

which leads to

$$\langle W\rangle \ \geq \Delta F - k_B T \ \langle I\rangle. \tag{3}$$

Equation (2) and (3) generalize the second law of thermodynamics taking into account the amount of information introduced into the system [22, 9]. Indeed Eq. (3) indicates that thank to information the work performed on the system to drive it between an initial and a final equilibrium states can be smaller than the free energy difference between the two states. Equation (2) has been directly tested in a single electron transistor [23].

2.1.2. *The autonomous Maxwell demon improves cooling.* An autonomous Maxwell demon using a local feedback mechanism which allows an efficient cooling of the system [24, 25]. The device, whose principle is sketched in Fig. 5a), is composed by a SET (Single Electron Transitor) formed by a small normal metallic island connected to two normal metallic leads by tunnel junctions, which permit electron transport between the leads and the island. The SET is biased by a potential V and a gate voltage V_g, applied to the island via a capacitance, controls the current I_e flowing through the SET. The island is coupled capacitively with a single electron box which acts as a demon which detects the presence of an electron in the island and applies a feedback. Specifically when an electron tunnels to the island, the demon traps it with a positive charge (panels 1 and 2). Conversely, when an electron leaves the island, the demon applies a negative charge to repel further electrons that would enter the island (panels 3 and 4). This effect is obtained by designing the electrodes of the demon in such a way that when an electron enters the island from a source electrode, an electron tunnels out of the demon island as a response, exploiting the mutual Coulomb repulsion between the two electrons. Similarly, when an electron enters to the drain electrode from the system island, an electron tunnels back to the demon island, attracted by the overall positive charge. The cycle of these interactions between the two devices realizes the autonomous demon, which allows the cooling of the leads. In the experimental realization presented in [24], the leads and the demon were thermally insulated, and the measurements of their temperatures is used to characterize the effect of the demon on the device operation. In Fig. 5b) we plot the variation of the leads temperatures as a function of $n_g \propto V_g$ when the demon acts on the system. We clearly see that around $n_g = 1/2$ the two leads are both cooled of $1mK$ at a mean temperature of $50mK$. This occurs because the tunneling electrons have to take the energy from the thermal energy of the leads, which, being thermally isolated, cool down. This increases the rate at which electrons tunnel against Coulomb repulsion, giving rise to increased cooling power. At the same time the demon increases its temperature because it has to dissipate energy in order to processes information, as discussed in Ref. [26]. Thus the total (system+demon) energy production is positive. The coupling of the demon with the SET can be controlled by a second gate which acts on the single electron box. In Fig. 5c) we plot the

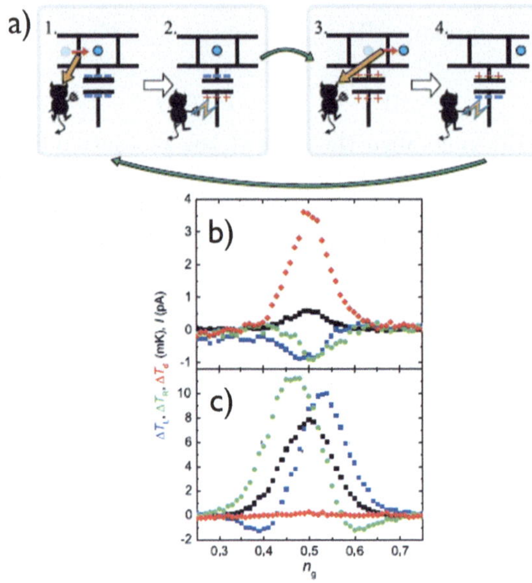

FIGURE 5. **a)** Principle of the experimental realization of the autonomous Maxwell demon. The horizontal top row schematizes a Single Electron Transistor. Electrons (blue circle) can tunnel inside the central island from the left wall and outside from the right wall. The demon watches at the state of the island and it applies a positive charge to attract the electrons when they tunnel inside and they repels them when they tunnel outside. The systems cools because of the energy released toward the heat bath by the tunneling events and the presence of the demon makes the cooling processes more efficient. The energy variation of the processes is negative because of the information introduced by the demon. **b)** The measured temperature variations of the left (bleu line) and right (green line) leads as a function of the external control parameter n_g when the demon is active and the bath temperature is $50mK$. We see that at the optimum value $n_g = 1/2$ both leads are cooled of about $1mK$ and the current I_e flowing through the SET (black line) has a maximum. At the same time in order to processes information the temperature of the demon (red line) increases of a few mK. **c)** The same parameter of the panel b) are measured when the demon is not active. We see that the demon temperature does not change, whereas both leads are now heated by the current I_e (adapted from Ref. [24]).

measured temperatures when the demon has been switched off. We clearly see that in such a case the demon temperature does not change and the two electrodes are heating up because of the current flow. This is the only example which shows that under specific conditions an autonomous local Maxwell demon, which does not use the external feedback, can be realized.

2.2. Experiments on Landauer's Principle

The experiments in the last section show that one can extract work from information. In the rest of this section we will discuss the reverse process, i.e., the energy needed to erase information (see section 1.2). Landauer's original thought experi-

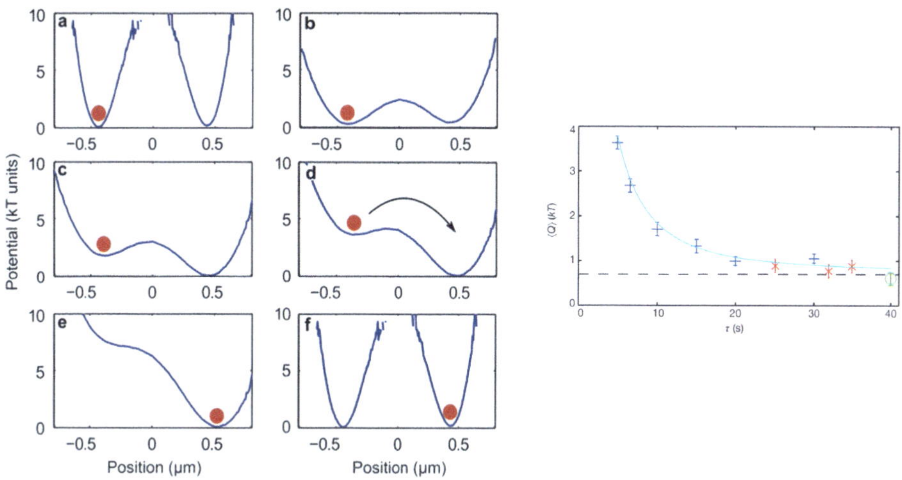

FIGURE 6. Experimental verification of Landauer's erasure principle. A colloidal particle is initially confined in one of two wells of a double-well potential with probability one-half. This configuration stores one bit of information. By modulating the height of the barrier and applying a tilt, the particle can be brought to one of the wells with probability one, irrespective of the initial position. This final configuration corresponds to zero bit of information. In the limit of long erasure cycles, the heat dissipated during the erasure process can approach, but not exceed, the Landauer bound indicated by the dashed line. A short description of the experiment can be found in Appendix B (adapted from Ref. [3]) .

ment has been realized for the first time in a real system in 2011 using a colloidal Brownian particle in a fluid trapped in a double-well potential produced by two strongly focused laser beams [3, 4, 5] (see also Appendix B). This system has two distinct states (particle in the right or left well) and may thus be used to store one bit of information. The erasure principle has been verified by implementing a protocol proposed by Bennett and illustrated in Fig. 6. At the beginning of the erasure process, the colloidal particle may be either in the left or right well with equal probability of one half. The erasure protocol is composed of the following steps: 1) the barrier height is first decreased by varying the laser intensity, 2) the particle is then pushed to the right by gently inclining the potential and 3) the potential is brought back to its initial shape. At the end of the process, the particle is in the right well with unit probability, irrespective of its departure position. As in the previous experiment, the position of the particle is recorded with the help of a camera. For a full erasure cycle, the average heat dissipated into the environment is equal to the average work needed to modulate the form of the double-well potential. This quantity was evaluated from the measured trajectory and shown to be always larger than the Landauer bound which is asymptotically approaches in the limit of long erasure times. However, in order to reach the bound, the protocol must be accurately chosen because as discussed in Ref. [3] and shown experimentally [27] there are protocols that are intrinsically irreversible no matter how slow are performed. The way in which a protocol can be optimized has been theoretically solved in Ref. [28] but the optimal protocol is not often easy to apply in an experiment

2.3. Other Experiments on the Physics of Information

By having successfully turned gedanken into real experiments, the above four seminal examples provide a firm empirical foundation to the physics of information and the intimate connection existing between information and energy. This connection is reinforced by the relationship between the generalized Jarzinsky equality [29] and the Landauer bound which has been proved and tested on experimental data in Ref. [4] and shortly summarized in the appendix A of this chapter .

A number of additional experiments have verified the erasure principle in various systems [30, 31, 32, 33, 34, 35, 36]. The latter include an electrical RC circuit [30] and a feedback trap [32, 33]. In addition, Ref. [34] has studied the symmetry breaking, induced in the probability distribution of the position of a Brownian particle, by commuting the trapping potential from a single to a double well potential. The authors measured the time evolution of the system entropy and showed how to produce work from information. Finally, experiments on the Landauer bound have been performed in nano devices, most notably using a single electron box [31] and nanomagnets [35, 36]. These experiments open the way to insightful applications for future developments of information technology.

3. Extensions to the Quantum Regime

3.1. Experiments on Quantum Maxwell's Demon

The experimental investigation of the physics of information has lately been extended to the quantum regime. The group of Roberto Serra in Sao Paulo has successfully realized a quantum Maxwell demon in a Nuclear Magnetic Resonance

(NMR) setup [37]. The demon was implemented as a spin-1/2 quantum memory that acquires information about another spin-1/2 system and employs it to control its dynamics. Using a coherent measured-based feedback protocol, the demon was shown to rectify the nonequilibrium entropy production due to quantum fluctuations and produce useful work. Concretely, the demon gained information about the system via a complete projective measurement. Based on the outcome of this measurement, a controlled evolution was applied to the system to balance the entropy production. Using quantum state tomography to reconstruct the density matrix ρ of the system at all times, the produced average work $\langle W \rangle$, or equivalently the mean entropy production $\langle \Sigma \rangle = \beta(\langle W \rangle - \Delta F)$, was shown to be bounded by the information gain, $\langle \Sigma \rangle \leq I_{\text{gain}}$. The latter quantifies the average information that the demon obtains by reading the outcomes of the measurement and is defined as $I_{\text{gain}} = S(\rho) - \sum_i p_i S(\rho_i)$, where ρ_i is the state after a measurement which occurs with probability p_i (see Fig. 7).

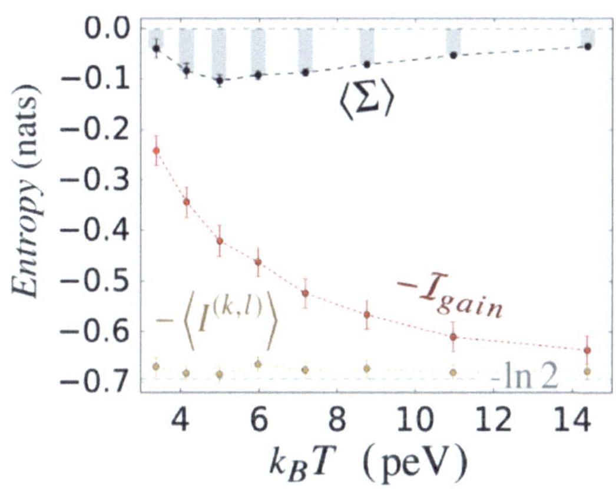

FIGURE 7. Thermodynamics of a quantum Maxwell demon. Verification of the second law for the nonequilibrium mean entropy production, $\langle \Sigma \rangle = \beta(\langle W \rangle - \Delta F) \leq I_{\text{gain}}$, in the presence of quantum feedback as a function of temperature. The parameter I_{gain} quantifies the information gained through the measurement (adapted from Ref. [37]).

More recently, a quantum Maxwell demon has been implemented in a circuit QED system [38]. Here, the demon was a microwave cavity that encodes quantum information about a superconducting qubit and converts that information into work by powering up a propagating microwave pulse by stimulated emission. The power extracted from the system was directly accessed by measuring the difference between incoming and outcoming photons of the cavity. Using full tomography of the system, the entropy remaining in the demon's memory was further quantified and was shown to be always higher that the system entropy decrease, in agreement with the second law.

In addition in a quantum demon setting a multi-photon optical interferometer allowed the measure of the extractable work which was used as a thermodynamic separability criterion to assess the entanglement of two-qubit and three-qubit systems [39]. An experimental analysis of two-qubit Bell states and three-qubit GHZ and W states has confirmed that more work can be extracted from an entangled state than from a separable state. Bounds on the extractable work can therefore be employed as a useful thermodynamic entanglement witness.

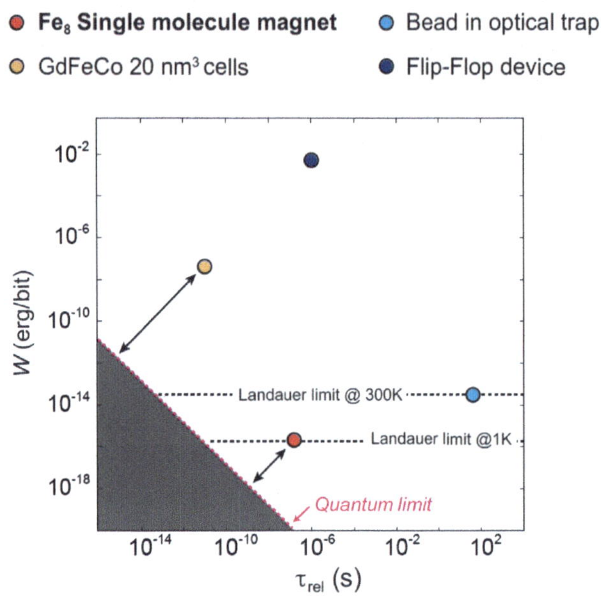

FIGURE 8. Energy-time cost of erasure. The diagram shows the product of the energy and the time needed for erasure, $W \cdot \tau_{\text{rel}}$, for various systems. The quantum limit is given by the Heisenberg uncertainty relation, $E \cdot \Delta t \geq \pi\hbar/2$. The Fe_8 molecule is currently the closest to the quantum limit (red dot) (adapted from Ref. [40]).

3.2. Experiments on Quantum Landauer's Principle

Erasure of information encoded in quantum states has been first theoretically considered by Lubkin [41] and Vedral [42] (see also Ref. [12]). An experimental verification of the Landauer principle in a quantum setting has been recently reported using a molecular nanomagnet at a temperature of 1K [40]. One bit of information was initially stored in a double-well potential of collective giant spin $S_z = \pm 10$ of a Fe_8 molecule. Work for the application of the tilt induced by a transverse magnetic field was determined via measurements of the magnetic susceptibility. Contrary to classical erasure which is achieved by decreasing the barrier height, here erasure was promoted by a thermally activated quantum tunnelling process. As a result, full erasure can be achieved much faster than in the classical regime. Using the product of the erasure work and the relaxation time, $W \cdot \tau_{\rm rel}$, as a figure of merit for the energy-time cost of information erasure, this experiment has reached the lower value to date with $W \cdot \tau_{\rm rel} \simeq 2 \cdot 10^{-23}$ erg/bit. As compared to 10^{-12} erg/bit. s for the classical experiment with the colloidal particle [3]. This puts the experiment close to the fundamental limit imposed by the Heisenberg uncertainty relation (see Fig. 8).

4. Applications

Landauer's principle applies not only to information erasure but also to all logically irreversible devices that possess more outputs than inputs. Thus, any Boolean gate operation that maps several input states onto the same output state, such as AND, NAND, and OR, have several states which are logically irreversible and will lead to the dissipation of an amount of heat of $k_B T \ln(2)$ per processed bit, akin to the erasure process. As a result, Landauer's principle has important technological consequences. Heating laptops are nowadays becoming part of everyday experience. Heat production in microprocessors used in modern computers is known to be a major factor hindering their miniaturization, as it gets more and more difficult to evacuate excess heat when size, and thus surface, is reduced. While the overall heat dissipated in microchips is steadily decreasing, it still several orders of magnitude larger than the Landauer limit. However, the switching energy of a CMOS/FET transistor is predicted to reach the Landauer bound by 2035, indicating that engineers will soon face a fundamental physical limitation imposed by the second law of thermodynamics [43, 44]. This is remarkable as $k_B T \ln(2)$ is about 3.10^{-21} Joule at room temperature and hence 22 orders of magnitude smaller than typical energy dissipated on our macroscopic scale. Recently, an experiment has demonstrated that Maxwell's demon can generate electric current and power by rectifying individual randomly moving electrons in small transistors [45].

Man-made computers are not the only existing information processing devices. Scientists have long realized that living biological cells can be viewed as biochemical information processors that may even outperform our current technology [46]. Cells are, for example, able to reproduce and create copies of themselves, acquire and process information coming from external stimuli, as well as communicate and exchange information with other cells. Recently, Landauer's principle has been employed to evaluate the energetic cost of a living cell computing the steady-state concentration of a chemical ligand in its surrounding environment [47]; it has been argued that it sets strong constraints on the design of cellular computing networks, as there is a tradeoff between the information processing capability of such a network and its energetic cost. Another important problem is the investigation of ultrasensitive switches in molecular biology. A concrete example is the flagellar motor of E. coli bacteria that switches from clockwise to counterclockwise rotation depending on the intracellular concentration of a regulator protein. Switching mechanisms are highly complex and not fully understood. A mathematical framework that models the sensing of the protein concentration by the flagellar motor as a Maxwell demon has been successfully developed to calculate the rate of energy consumption needed to both sense and switch, and provide a quantitative description of the switching statistics [48]. More recent work has focused on the efficiency of cellular information processing [49], biochemical signal transduction [50], as well as on cost and precision of Brownian clocks [51] and computational copying in biochemical systems [52].

Maxwell's demon is therefore still vibrant 150 years after its inception. Together with Landauer's principle, he continues to play a prominent role in modern research as illustrated by the last examples. Having only very recently become an experimental science, information physics appears to have a promising future ahead.

Appendix A. Stochastic Thermodynamics and Information Energy Cost

When the size of a system is reduced the role of fluctuations (either quantum or thermal) increases. Thus thermodynamic quantities such as internal energy, work, heat and entropy cannot be characterized only by their mean values but also their fluctuations and probability distributions become relevant and useful to make predictions on a small system. Let us consider a simple example such as the motion of a Brownian particle subjected to a constant external force. Because of thermal fluctuations, the work performed on the particle by this force per unit time, i.e., the injected power, fluctuates and the smaller the force, the larger is the importance of power fluctuations [53, 54, 55]. The goal of stochastic thermodynamics is just that of studying the statistical properties of the above mentioned fluctuating thermodynamic quantities in systems driven out of equilibrium by external forces, temperature differences and chemical reactions. For this reason it has received in the last twenty years an increasing interest for its applications in microscopic devices, biological systems and for its connections with information theory[53, 54, 55].

Specifically it can be shown that the fluctuations on a time scale τ of the internal energy ΔU_τ, the work W_τ and the heat Q_τ are related by a first principle like equation, i.e.,

$$\Delta U_\tau = U(t+\tau) - U(t) = \tilde{W}_\tau - Q_\tau \tag{4}$$

at any time t.

Furthermore the statistical properties of energy and entropy fluctuations are constrained by fluctuations theorems which impose bounds on their probability distributions (for more details see ref.[53, 54, 55]). We summarize in the next section one of them which is can be related to information and to Landauer's bound.

A.1. Estimate the Free Energy Difference from Work Fluctuations

In 1997 [56, 57] Jarzynski derived an equality which relates the free energy difference of a system in contact with a heat reservoir to the pdf of the work performed on the system to drive it from A to B along any path γ in the system parameter space. Specifically, when a system parameter λ is varied from time $t = 0$ to $t = t_s$, Jarzynski defines for one realization of the "switching process" from A to B the work performed on the system as

$$W_{\mathrm{st}} = \int_0^{t_s} \dot{\lambda}\, \frac{\partial H_\lambda[z(t)]}{\partial \lambda} \mathrm{d}t, \tag{5}$$

where z denotes the phase-space point of the system and H_λ its λ-parametrized Hamiltonian [1]. One can consider an ensemble of realizations of this "switching process" with initial conditions all starting in the same initial equilibrium state. Then W_{st} may be computed for each trajectory in the ensemble. The Jarzynski equality states that [56, 57]

$$\exp(-\beta \Delta F) = \langle \exp(-\beta W_{\mathrm{st}}) \rangle, \tag{6}$$

where $\langle \cdot \rangle$ denotes the ensemble average, $\beta^{-1} = k_B T$ with k_B the Boltzmann constant and T the temperature. In other words $\langle \exp[-\beta W_{\mathrm{diss}}] \rangle = 1$, since we can always write $W_{\mathrm{st}} = \Delta F + W_{\mathrm{diss}}$ where W_{diss} is the dissipated work. Thus it is easy to see that there must exist some paths γ such that $W_{\mathrm{diss}} \leq 0$. Moreover, the inequality $\langle \exp x \rangle \geq \exp\langle x \rangle$ allows us to recover the second principle, namely $\langle W_{\mathrm{diss}} \rangle \geq 0$, i.e., $\langle W_{\mathrm{st}} \rangle \geq \Delta F$.

A.2. Landauer Bound and the Jarzynski Equality

We discuss in this appendix the strong relationship between the Jarzynski equality and the Landauer's bound. In Box 1 we presented the Landauer's principle as related to the system entropy. Let us consider as a specific example the experiment on the colloidal particle described in section 2.2 [4]. In the memory erasure procedure which forces the system in the state 0, the entropy difference between the final and initial state is $\Delta S = -k_{\mathrm{B}} \ln(2)$. In contrast the internal internal energy is unchanged by the protocol. Thus it is natural to await $\Delta F = k_{\mathrm{B}} T \ln(2)$. However the ΔF that appears in the Jarzynski equality is the difference between the free energy of the system in the initial state (which is at equilibrium) and the equilibrium state corresponding to the final value of the control parameter: $F(\lambda(\tau)) - F(\lambda(0))$. Since the height of the barrier is always finite there is no change in the equilibrium free energy of the system between the beginning and the end of the procedure. Then $\Delta F = 0$, which implies $\left\langle e^{-\beta W_{\mathrm{st}}} \right\rangle = 1$. Thus it seems that there is a problem between the Landauer principle (see Box 1) and the Jarzynski equality of Eq. 6.

---

[1] This is a more general definition of work and it coincides with the standard one only if λ is a displacement (for more details see ref.[55]

Nevertheless Vaikuntanathan and Jarzynski [29] have shown that when there is a difference between the actual state of the system (described by the phase-space density ρ_t), and the equilibrium state (described by ρ_t^{eq}), the Jarzynski equality can be modified:

$$\left\langle e^{-\beta W_{\mathrm{st}}(t)} \right\rangle_{(x,t)} = \frac{\rho^{\mathrm{eq}}(x, \lambda(t))}{\rho(x,t)} e^{-\beta \Delta F(t)}, \tag{7}$$

where $\langle . \rangle_{(x,t)}$ is the mean on all the trajectories that pass through x at time t.

In the experiment presented in section 2.2, the selection of the trajectories where the information is actually erased, corresponds to fix x to the chosen final well at the time $t = \tau$. It follows that $\rho(0, \tau)$ is the probability of finding the particle in the targeted state 0 at the time τ. Indeed because of the very low energy measured in the protocol thermal fluctuations play a role and the particle can be found in the wrong well at time τ, i.e., the proportion of success P_S of the procedure is equal to $\rho(0, \tau)$. In contrat the equilibrium distribution is $\rho^{\mathrm{eq}}(0, \lambda(\tau)) = 1/2$. Then:

$$\left\langle e^{-\beta W_{\mathrm{st}}(\tau)} \right\rangle_{\to 0} = \frac{1/2}{P_S}. \tag{8}$$

Similarly for the trajectories that end the procedure in the wrong well (i.e., state 1) we have:

$$\left\langle e^{-\beta W_{\mathrm{st}}(\tau)} \right\rangle_{\to 1} = \frac{1/2}{1 - P_S}. \tag{9}$$

Taking into account the Jensen inequality, i.e., $\langle e^{-x} \rangle \geq e^{-\langle x \rangle}$, we find that equations 8 and 9 imply:

$$\begin{aligned} \langle W_{\mathrm{st}} \rangle_{\to 0} &\geq k_{\mathrm{B}} T \left[\ln(2) + \ln(P_S) \right], \\ \langle W_{\mathrm{st}} \rangle_{\to 1} &\geq k_{\mathrm{B}} T \left[\ln(2) + \ln(1 - P_S) \right]. \end{aligned} \tag{10}$$

Notice that the mean work dissipated to realize the procedure is simply:

$$\langle W_{\mathrm{st}} \rangle = P_S \times \langle W_{\mathrm{st}} \rangle_{\to 0} + (1 - P_S) \times \langle W_{\mathrm{st}} \rangle_{\to 1}, \tag{11}$$

where $\langle . \rangle$ is the mean on all trajectories. Then using the previous inequalities it follows:

$$\langle W_{\mathrm{st}} \rangle \geq k_{\mathrm{B}} T \left[\ln(2) + P_S \ln(P_S) + (1 - P_S) \ln(1 - P_S) \right], \tag{12}$$

which is indeed the generalization of the Landauer's bound for $P_S < 1$. In the limit case where $P_S \to 1$, we have:

$$\left\langle e^{-\beta W_{\mathrm{st}}} \right\rangle_{\to 0} = 1/2. \tag{13}$$

Since this result remains approximatively verified for proportions of success close enough to 100%, it explains why in the experiment we find $\Delta F_{\mathrm{eff}} \approx k_{\mathrm{B}} T \ln(2)$.

This result is useful because it strongly binds the generalized Jarzynski equality (a thermodynamic relation) to Landauer's bound.

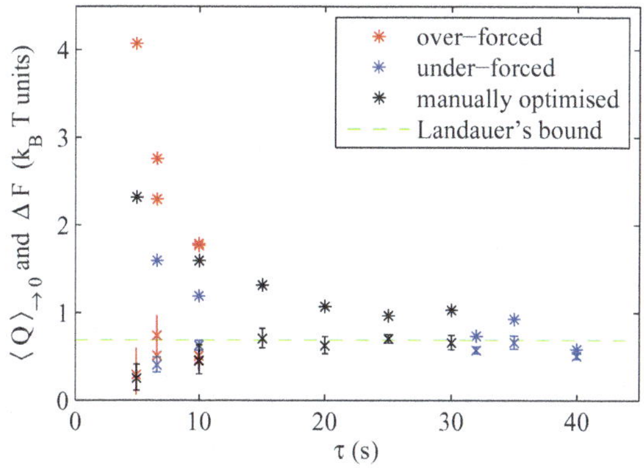

FIGURE 9. Mean dissipated heat ($*$) and effective free energy difference (Eq. 14) ($\times$) for several procedures, with fixed τ and different values of $F_{\max}$. The red points have a force too high, and a $P_S \geq 99\,\%$. The blue points have a force too low and $91\,\% \leq P_S < 95\,\%$ (except the last point which has $P_S \approx 80\,\%$). The black points are considered to be optimised and have $95\,\% \leq P_S < 99\,\%$(adapted from Ref. [5]).

A.2.1. Experimental test of the generalized Jarzynski equality. The theoretical predictions of the previous section have been checked in the experiment on the colloidal particle presented in Section 2.2 and in Fig.6. In ref. [4] the authors experimentally compute ΔF_{eff} which is the logarithm of the exponential average of the dissipated heat for trajectories ending in state 0:

$$\Delta F_{\text{eff}} = -\ln\left(\left\langle e^{-\beta W_{\text{st}}}, \right\rangle_{\to 0}\right) \tag{14}$$

where W_{st} in the experiment is equal to Q as explained in Appendix B.2 (Eq. 17).

Data are shown in Fig. 9. The error bars are estimated by computing the average on the data set with 10% of the points randomly excluded, and taking the maximal difference in the values observed by repeating this operation 1000 times. Except for the first points ($\tau = 5\,\text{s}$), the values of ΔF_{eff} are very close to $k_B T \ln 2$ for any value of the protocol duration τ, which is in agreement with equation 9, since P_S is close to $100\,\%$. Hence, we retrieve the Landauer's bound for the free-energy difference, for any duration of the information erasure procedure.

Note that this result is not in contradiction with the classical Jarzynski equality, because if we average over all the trajectories (and not only the ones where the information is erased), we should find $\left\langle e^{-\beta W_{\text{st}}}\right\rangle = 1$.

In ref.[4, 5] the authors looked at the two sub-procedures $1 \to 0$ and $0 \to 0$ separately, finding an excellent agreement with Eq. 12.

Appendix B. Set-up Used in the Experiment Presented in Section 2.2

B.1. The One-Bit Memory System

The one-bit memory system, is made of a double well potential where one particle is trapped by optical tweezers. If the particle is in the left-well the system is in the state "0", if the particle is in the right-well the system in the state "1". Full details of the experimental set-up can be found in ref.[5] and we summarize here the main features. We use a custom-built vertical optical tweezer made of an oil-immersion objective (63×, N.A. = 1.4) which focuses a laser beam (wavelength $\lambda = 1064$nm) to the diffraction limit for trapping glass beads (2μm in diameter) [5]. The beads are dispersed in bidistilled water in very small concentration. The suspension is introduced in a disk-shaped cell (18mm in diameter, 1mm in depth), a single bead is then trapped and moved away from the others. This step is quite important to avoid that during the measurement the trapped bead is perturbed by other Brownian particles. The position of the bead is tracked using a fast camera with a resolution of 108nm/pixels, which gives after treatment the position with an accuracy better than 10nm. The trajectories of the bead are sampled at 502Hz. The double-well potential is obtained by switching the laser at a rate of 10KHz between two points at a distance $d_f = 1.45\mu$m, kept fixed. The form of the potential, which is a function of d_f and the laser intensity I_L, can be determined in equilibrium by measuring the probability

$$P(x, I_L) = A \exp[-U_o(x, I_L)/(kT)] \tag{15}$$

of the position x of the bead, i.e., $U_o(x, I_L) = -kT \ln[P(x, I_L)/A]$ (see Fig. 6 a,b and f) The distribution $P(x, I_L)$ is estimated on about 10^6 samples. The measured $U_o(x, I_L)$ are plotted in Figs. 6 a,b and f) and can be fitted by an eighth order polynomial $U_o(x, I_L) = \sum_{n=0}^{8} u_n(I_L, d_f)x^n$. The distance between the two minima of the double-well potential is 0.9μm. The two wells are nearly symmetric with a maximum energy difference of $0.4kT$. The height of the barrier is modulated by varying the power of the laser from $I_L = 48mW$ (barrier height $> 8kT$) to $I_L = 15mW$ (barrier height $= 2.2kT$). In equilibrium for a barrier of $8kT$, the characteristic jumping time (Kramers time) between the two wells is about 3000s which is much longer than any experimental time.

The external tilt is created by displacing the cell with respect to the laser with a piezo, thus inducing a viscous flow. The viscous force is simply $F = -\gamma v$, where $\gamma = 1.89N.s.m^{-1}$ is the friction coefficient and v the velocity of the cell. In the erasure protocol, the amplitude of the viscous force is increased linearly during time τ, $F(t) = F_{max}\, t/\tau$. In Figs. 6 a,b and f) , we plot $U(x, t) = U_o(x, I_L) - F(t)\, x$ for $I_L = 15mW$ and for three different values of t. The typical erasure protocol is presented in Fig.10. The reinitialization procedure shown in this figure is necessary to displace the cell to its initial position but it does not contribute to the erasure process. Notice that, contrary to the useful erasure cycles, this reinitialization is performed when the barrier is high. Thus the bead remains always in the same

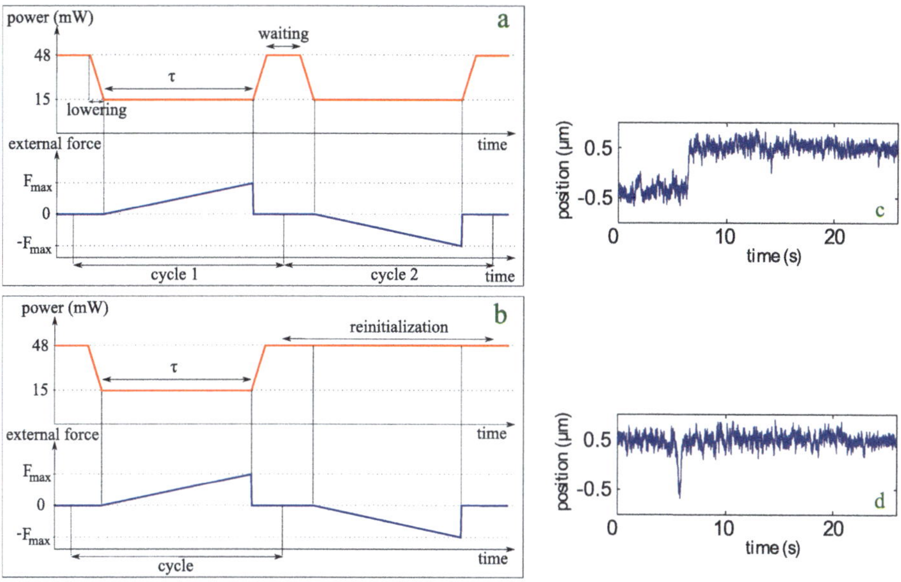

FIGURE 10. Erasure cycles and typical trajectories. a) Protocol used for the erasure cycles bringing the bead from left (state 0) to right (state 1), and vice versa. b) Protocol used to measure the heat for the cycles for which the bead does not change wells. The reinitialization is needed to restart the measurement, but is not a part of the erasure protocol. c) Example of a measured bead trajectory for the transition $0 \to 1$. d) Example of a bead trajectory for the transition $1 \to 1$.

well. Note that for the theoretical procedure, the system must be prepared in an equilibrium state with same probability to be in state 1 than in state 0. However, it is more convenient experimentally to have a procedure always starting in the same position. Therefore we separate the procedure in two sub-procedures: one where the bead starts in state 1 and is erased in state 0, and one where the bead starts in state 0 and is erased in state 0. The fact that the position of the bead at the beginning of each procedure is actually known is not a problem because this knowledge is not used by the erasure procedure. The important points are that there is as many procedures starting in state 0 than in state 1, and that the procedure is always the same regardless of the initial position of the bead. Examples of trajectories for the two sub-procedures 1 ? 0 and 0 ? 0 are shown in Fig.10.

A key characteristic of the erasure process is its success rate P_S, that is, the relative number of cycles bringing the bead in the expected well. Fig.11a) shows the dependence of the erasure rate on the tilt amplitude, F_{max}. For definiteness,

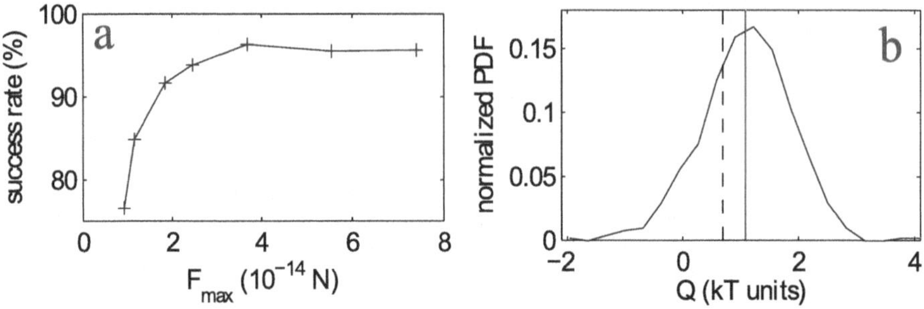

FIGURE 11. a) Success rate P_S of the erasure cycle as a function of the maximum tilt amplitude F_{max}. b) Heat distribution $P(Q)$ for transition $0 \to 1$ with $\tau = 25s$ and $F_{max} = 1.89 \times 10^{-14}$N. The solid vertical line indicates the mean dissipated heat, and the dashed vertical line marks the Landauer bound.

we have kept the product $F_{max}\tau$ constant. We observe that the erasure rate drops sharply at low amplitudes when the tilt force is too weak to push the bead over the barrier, as expected. For large values of F_{max}, the erasure rate saturates at around 95%. This saturation reflects the finite size of the barrier and the possible occurrence of spontaneous thermal activation into the wrong well. An example of a distribution of the dissipated heat for the transition $0 \to 1$ is displayed in Fig.11b). Owing to thermal fluctuations, the dissipated heat may be negative and maximum erasure below the Landauer limit may be achieved for individual realizations, but not on average [3, 5].

B.2. Heat Measurements

The heat dissipated by the tilt is

$$Q = \int_{T_{low}}^{T_{low}+\tau} dt\, F(t)\dot{x}(t) = \int_{T_{low}}^{T_{low}+\tau} dt\, F_{max}\,\frac{t}{\tau}\dot{x}(t), \tag{16}$$

where T_{low} is the time at which the barrier is reduced to its minimum value. The velocity is computed using the discretization, $\dot{x}(t+\Delta t/2) \simeq [x(t+\Delta t) - x(t)]/\Delta t$. To characterize the approach to the Landauer limit we note that, in the quasi-static limit, the mean work $< W >$ can be expressed in terms of the free energy difference ΔF as $< W > \simeq \Delta F + C/\tau$ [53](see Appendix A). According to the first law of thermodynamics, $< \Delta U > = < W > - < Q > = 0$ for a cycle. As a result, $\Delta F = -T\Delta S$ and $< Q > = < W > \simeq kT \ln 2 + C/\tau$.

Finally we notice that in the used protocol the $W_{\rm st}$ defined in Eq. 5 is equal to Q. Indeed as $F(t = T_{\rm low}) = 0 = F(t = T_{\rm low} + \tau)$ it follows from an integration by parts that the stochastic work is equal to the heat dissipated by the force:

$$W_{\rm st} = \int_{T_{\rm low}}^{T_{\rm low}+\tau} -\dot{F}x\,{\rm d}t' = \int_{T_{\rm low}}^{T_{\rm low}+\tau} F\dot{x}\,{\rm d}t' = Q. \tag{17}$$

References

[1] S. Ciliberto and E. Lutz, The physics of information: From Maxwell to Landauer. In: C.S. Lent, A.O. Orlov, W. Porod, and G.L. Snider, editors, *Energy Limits in Computation*, chapter 5, pages 155–176. Springer, 2018.

[2] E. Lutz and S. Ciliberto, Information: From Maxwell's demon to Landauer's eraser. *Physics Today* **68** (9), 30 (2015).

[3] A. Bérut, A. Arakelyan, A. Petrosyan, S. Ciliberto, R. Dillenschneider, and E. Lutz, Experimental verification of Landauer's principle linking information and thermodynamics. *Nature* **483**, 187 (2012).

[4] A. Bérut, A. Petrosyan, and S. Ciliberto, Detailed Jarzynski equality applied to a logically irreversible procedure. *Europhys. Lett.* **103** (6), 60002 (2013).

[5] A. Bérut, A. Petrosyan, and S. Ciliberto, Information and thermodynamics: experimental verification of Landauer's erasure principle. *J. Stat. Mech.* P06015 (2015).

[6] R. Landauer, Information is physical. *Physics Today.* **44** (5), (1991).

[7] W. Weaver and C.E. Shannon, *The Mathematical Theory of Communication.* Univ. Illinois Press, Urbana, 2010.

[8] R. Clausius, *The Mechanical Theory of Heat.* McMillan, London, 1879.

[9] J.M.R. Parrondo, J. M. Horowitz, and T. Sagawa, Thermodynamics of information. *Nature Physics* **11**, 131 (2015).

[10] H.S. Leff and A.F. Rex, eds., *Maxwell's Demon: Entropy, Classical and Quantum Information, Computing.* Institute of Physics, Philadelphia, 2003.

[11] M.B. Plenio and V. Vitelli, The physics of forgetting: Landauer's erasure principle and information theory. *Contemp. Phys.* **42**, 25 (2010).

[12] K.K. Maruyama, F. Nori, and V. Vedral, The physics of Maxwell's demon and information. *Rev. Mod. Phys.* **81**, 1 (2009).

[13] L. Szilard, On the minimization of entropy in a thermodynamic sytem with interferences of intelligent beings. *Z. Physik* **53**, 840 (1929).

[14] C.H. Bennett, The thermodynamcis of computation-a review. *Int. J. Theor. Phys.* **21**, 905 (1982).

[15] H.S. Leff and A.F. Rex, eds., *Foundations of Statistical Mechanics.* Pergamon, Oxford, 1970.

[16] L. Brillouin, *Science and Information Theory.* Academic Press, Waltham, MA, 1956.

[17] M.G. Raizen, Comprehensive control of atomic motion. *Science* **324**, 1403 (2009).

[18] G.N. Price, S.T. Bannerman, K. Viering, and E. Narevicius, Single-photon atomic cooling. *Phys. Rev. Lett.* **100**, 093004 (2008).

[19] J.J. Thorn, E.A. Schoene, T. Li, and D.A. Steck, Experimental realization of an optical one-way barrier for neutral atoms. *Phys. Rev. Lett.* **100**, 240407 (2008).

[20] S. Toyabe, T. Sagawa, M. Ueda, M. Muneyuki, and M. Sano, Experimental demonstration of information-to-energy conversion and validation of the generalized Jarzynski equality. *Nature Phys.* **6**, 988 (2010).

[21] T. Sagawa and M. Ueda, Generalized Jarzynski equality under nonequilibrium feedback control. *Phys. Rev. Lett.* **104**, 090602 (2010).

[22] T. Sagawa and M. Ueda, Minimum energy cost for thermodynamic information processing: Measurement and information erasure. *Phys. Rev. Lett.* **102**, 250602 (2009).

[23] J.V. Koski, V.F. Maisi, T. Sagawa, and J.P. Pekola, Experimental observation of the role of mutual information in the nonequilibrium dynamics of a Maxwell's demon. *Phys. Rev. Lett.* **113**, 030601 (2014a).

[24] J.V. Koski, A. Kutvonen, I.M. Khaymovich, T. Ala-Nissila, and J.P. Pekola, On-chip Maxwell's demon as an information-powered refrigerator. *Phys. Rev. Lett.* **115**, 260602 (2015).

[25] J.V. Koski and J.P. Pekola, Maxwell's demons realized in electronic circuits. *C.R. Physique* **17**, 1130 (2016).

[26] J.M. Horowitz and M. Esposito, Thermodynamics with continuous information flow. *Phys. Rev. X* **4**, 031015 (2014).

[27] M. Gavrilov and J. Bechhoefer, Arbitrarily slow, non-quasistatic, isothermal transformations. *Europhys. Lett.* **114**, 50002 (2016a).

[28] E. Aurell, K. Gawedzki, C. Meja-Monasterio, R. Mohayaee, and P. Muratore-Ginanneschi, Refined second law of thermodynamics for fast random processes. *J. Stat. Phys.* **147**, 487 (2012).

[29] S. Vaikuntanathan and C. Jarzynski, Dissipation and lag in irreversible processes. *Europhys. Lett.* **87**, 60005 (2009).

[30] A.O. Orlov, C.S. Lent, C.C. Thorpe, G.P. Boechler, and G.L. Snider, Experimental test of Landauer's principle at the sub- k_BT level. *Japanese Journal of Applied Physics* **51** (6S), 06FE10 (2012).

[31] J.V. Koski, V.F. Maisi, J.P. Pekola, and D.V. Averin, Experimental realization of a Szilard engine with a single electron. *Proc. Natl. Acad. Sci.* **111**, 13786 (2014b).

[32] Y. Jun, M. Gavrilov, and J. Bechhoefer, High-precision test of Landauer's principle in a feedback trap. *Phys. Rev. Lett.* **113**, 190601 (2014).

[33] M. Gavrilov and J. Bechhoefer, Erasure without work in an asymmetric, double-well potential. *Phys. Rev. Lett.* **117**, 200601 (2016b).

[34] E. Roldàn, I.A. Martinez, J.M.R. Parrondo, and D. Petrov, Universal features in the energetics of symmetry breaking. *Nature Physics* **10**, 457 (2014).

[35] L. Martini, M. Pancaldi, M. Madami, P. Vavassori, G. Gubbiotti, S. Tacchi, F. Hartmann, M. Emmerling, S. Höfling, L. Worschech, and G. Carlotti, Experimental and theoretical analysis of Landauer erasure in nano-magnetic switches of different sizes. *Nano Energy* **19** (Supplement C), 108 (2016).

[36] J. Hong, B. Lambson, S. Dhuey, and J. Bokor, Experimental test of Landauer's principle in single-bit operations on nanomagnetic memory bits. *Sci. Adv.* **2**, e1501492 (2016).

[37] P.A. Camati, J.P.S. Peterson, T.B. Batalhão, K. Micadei, A.M. Souza, R.S. Sarthour, I.S. Oliveira, and R.M. Serra, Experimental rectification of entropy production by Maxwell's demon in a quantum system. *Phys. Rev. Lett.* **117**, 240502 (2016).

[38] N. Cottet, S. Jezouin, L. Bretheau, P. Campagne-Ibarcq, Q. Ficheux, J. Anders, A. Auffèves, R. Azouit, P. Rouchon, and B. Huard, Observing a quantum Maxwell demon at work. *Proceedings of the National Academy of Sciences* **114** (29), 7561 (2017).

[39] M.A. Ciampini, L. Mancino, A. Orieux, C. Vigliar, P. Mataloni, M. Paternostro, and M. Barbieri, Experimental extractable work-based multipartite separability criteria. *npj Quantum Information* **3** (1), 10 (2017).

[40] R. Gaudenzi, E. Burzurí, S. Maegawa, H.S.J. van der Zant, F. Luis, Quantum Landauer erasure with a molecular nanomagnet (Supplementary Information). *Nature Physics* **14**, 565 (2018).

[41] E. Lubkin, Keeping the entropy of measurement: Szilard revisited. *International Journal of Theoretical Physics* **26** (6), 523 (1987).

[42] V. Vedral, Landauer's erasure, error correction and entanglement. *Proceedings of the Royal Society of London A: Mathematical, Physical and Engineering Sciences* **456** (1996), 969 (2000).

[43] M.P. Frank, The physical limits of computing. *Comput. Sci. Eng.* **4**, 16 (2002).

[44] E. Pop, Energy dissipation and transport in nanoscale devices. *Nano Res.* **3**, 147 (2010).

[45] K. Chida, S. Desai, K. Nishiguchi, and A. Fujiwara, Power generator driven by Maxwell's demon. *Nat. Comm* **8**, 15310 (2017).

[46] D. Bray, Energetic costs of cellular computation. *Nature* **376**, 307 (1995).

[47] P. Mehta and D.J. Schwab, Energetic costs of cellular computation. *Proc. Natl. Acad. Sci.* **109**, 17978 (2012).

[48] Y. Tu, The nonequilibrium mechanism for ultrasensitivity in a biological switch: Sensing by Maxwell's demons. *Proc. Natl. Acad. Sci.* **105**, 11737 (2008).

[49] A.C. Barato, D. Hartich, and U. Seifert, Efficiency of cellular information processing. *New J. Phys.* **16**, 103024 (2014).

[50] S. Ito and T. Sagawa, Maxwell's demon in biochemical signal transduction with feedback loop. *Nat. Comm.* **6**, 7498 (2015).

[51] A.C. Barato and U. Seifert, Cost and precision of brownian clocks. *Phys. Rev. X* **6**, 041053 (2016).

[52] T.E. Ouldridge, C.C. Govern, and P. Rein ten Wolde, Thermodynamics of computational copying in biochemical systems. *Phys. Rev. X* **7**, 021004 (2017).

[53] K. Sekimoto, *Stochastic Energetics*. Volume 799 of *Lecture Notes in Physics*. Springer, 2010.

[54] U. Seifert, Stochastic thermodynamics, fluctuation theorems and molecular machines. *Reports on Progress in Physics* **75** (12), 126001 (2012).

[55] S. Ciliberto, Experiments in stochastic thermodynamics: Short history and perspectives. *Phys. Rev. X* **7**, 021051 (2017).

[56] C. Jarzynski, Nonequilibrium equality for free energy differences. *Phys. Rev. Lett.* **78**, 2690 (1997a).

[57] C. Jarzynski, Equilibrium free-energy differences from nonequilibrium measurements: A master-equation approach. *Phys. Rev. E* **56**, 5018 (1997b).

Sergio Ciliberto
Université de Lyon
CNRS
Laboratoire de Physique
École Normale Supérieure de Lyon (UMR5672)
46 Allée d'Italie
69364 Lyon Cedex 07
France
e-mail: sergio.ciliberto@ens-lyon.fr

Information Theory, 113–209

Poincaré Seminar 2018

Verification of Quantum Computation: An Overview of Existing Approaches

Alexandru Gheorghiu, Theodoros Kapourniotis and Elham Kashefi

Abstract. Quantum computers promise to efficiently solve not only problems believed to be intractable for classical computers, but also problems for which verifying the solution is also considered intractable. This raises the question of how one can check whether quantum computers are indeed producing correct results. This task, known as *quantum verification*, has been highlighted as a significant challenge on the road to scalable quantum computing technology. We review the most significant approaches to quantum verification and compare them in terms of structure, complexity and required resources. We also comment on the use of cryptographic techniques which, for many of the presented protocols, has proven extremely useful in performing verification. Finally, we discuss issues related to fault tolerance, experimental implementations and the outlook for future protocols.

1. Introduction

Quantum computation is the subject of intense research due to the potential of quantum computers to efficiently solve problems which are believed to be intractable for classical computers. The current focus of experiments, aiming to realize scalable quantum computation, is to demonstrate a *quantum computational advantage*. In other words, this means performing a quantum computation in order to solve a problem which is proven to be classically intractable, based on plausible complexity-theoretic assumptions. Examples of such problems, suitable for near-term experiments, include *boson sampling* [1], *instantaneous quantum polynomial time* (IQP) computations [2] and others [3–5]. The prospect of achieving these tasks has ignited a flurry of experimental efforts [6–9]. However, while demonstrating a quantum computational advantage is an important milestone towards scalable quantum computing, it also raises a significant challenge:

> *If a quantum experiment solves a problem which is proven to be intractable for classical computers, how can one verify the outcome of the experiment?*

The first researcher who formalised the above "paradox" as a *complexity theoretic* question was Gottesman, in a 2004 conference [10]. It was then promoted, in 2007, as a complexity challenge by Aaronson who asked: "*If a quantum computer can efficiently solve a problem, can it also efficiently convince an observer that the solution is correct? More formally, does every language in the class of quantumly tractable problems* (BQP) *admit an interactive proof where the prover is in* BQP *and the verifier is in the class of classically tractable problems* (BPP)*?*" [10]. Vazirani, then emphasized the importance of this question, not only from the perspective of complexity theory, but from a philosophical point of view [11]. In 2007, he raised the question of whether quantum mechanics is a *falsifiable theory*, and suggested that a computational approach could answer this question. This perspective was explored in depth by Aharonov and Vazirani in [12]. They argued that although many of the predictions of quantum mechanics have been experimentally verified to a remarkable precision, all of them involved systems of low complexity. In other words, they involved few particles or few degrees of freedom for the quantum mechanical system. But the same technique of "*predict and verify*" would quickly become infeasible for systems of even a few hundred interacting particles due to the exponential overhead in classically simulating quantum systems. And so what if, they ask, the predictions of quantum mechanics start to differ significantly from the real world in the high complexity regime? How would we be able to check this? Thus, the fundamental question is whether there exists a verification procedure for quantum mechanical predictions which is efficient for arbitrarily large systems.

In trying to answer this question we return to complexity theory. The primary complexity class that we are interested in is BQP, which, as mentioned above, is the class of problems that can be solved efficiently by a quantum computer. The analogous class for classical computers, with randomness, is denoted BPP. Finally, concerning verification, we have the class MA, which stands for Merlin-Arthur. This consists of problems whose solutions can be verified by a BPP machine when given a proof string, called a *witness*[1]. BPP is contained in BQP, since any problem which can be solved efficiently on a classical computer can also be solved efficiently on a quantum computer. Additionally BPP is contained in MA since any BPP problem admits a trivial empty witness. Both of these containments are believed to be strict, though this is still unproven.

What about the relationship between BQP and MA? Problems are known that are contained in both classes and are believed to be outside of BPP. One

[1] BPP and MA are simply the probabilistic versions of the more familiar classes P and NP. Under plausible derandomization assumptions, BPP = P and MA = NP [13].

such example is *factoring*. Shor's polynomial-time quantum algorithm for factoring demonstrates that the problem is in BQP [14]. Additionally, for any number to be factored, the witness simply consists of a list of its prime factors, thus showing that the problem is also in MA. In general, however, it is believed that BQP is *not* contained in MA. An example of a BQP problem not believed to be in MA is approximating a knot invariant called the *Jones polynomial* which has applications ranging from protein folding to Topological Quantum Field Theory (TQFT) [15–17]. The conjectured relationship between these complexity classes is illustrated in Figure 1.

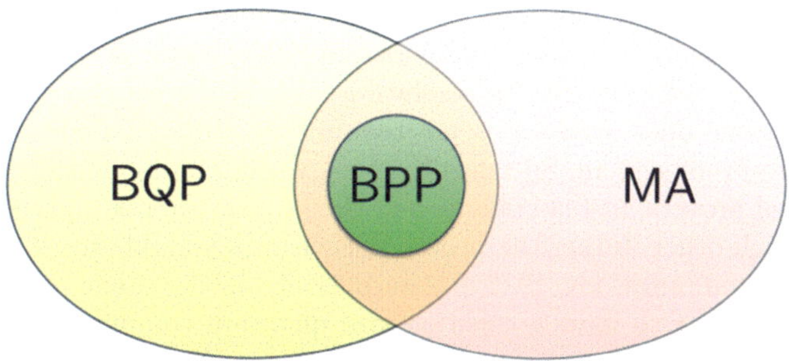

FIGURE 1. Suspected relationship between BQP and MA

What this tells us is that, very likely, there do not exist witnesses certifying the outcomes of general quantum experiments[2]. We therefore turn to a generalization of MA known as an *interactive-proof system*. This consists of two entities: a *verifier* and a *prover*. The verifier is a BPP machine, whereas the prover has unbounded computational power. Given a problem for which the verifier wants to check a reported solution, the verifier and the prover interact for a number of rounds which is polynomial in the size of the input to the problem. At the end of this interaction, the verifier should accept a valid solution with high probability and reject, with high probability, otherwise. The class of problems which admit such a protocol is denoted IP[3]. In contrast to MA, instead of having a single proof string for each problem, one has a transcript of back-and-forth communication between the verifier and the prover.

If we are willing to allow our notion of verification to include such interactive protocols, then one would like to know whether BQP is contained in IP. Unlike the relation between BQP and MA, it is, in fact, the case that BQP $\subset$ IP, which means

[2]Even if this were the case, i.e., BQP $\subseteq$ MA, for this to be useful in practice one would require that computing the witness can also be done in BQP. In fact, there are candidate problems known to be in both BQP and MA, for which computing the witness is believed to not be in BQP (a conjectured example is [18]).

[3]MA can be viewed as an interactive-proof system where only one message is sent from the prover (Merlin) to the verifier (Arthur).

that every problem which can be efficiently solved by a quantum computer admits an interactive-proof system. One would be tempted to think that this solves the question of verification, however, the situation is more subtle. Recall that in IP, the prover is computationally unbounded, whereas for our purposes we would require the prover to be restricted to BQP computations. Hence, the question that we would like answered and, arguably, the main open problem concerning quantum verification is the following:

Problem 1 (Verifiability of BQP computations). *Does every problem in* BQP *admit an interactive-proof system in which the prover is restricted to* BQP *computations?*

As mentioned, this complexity theoretic formulation of the problem was considered by Gottesman, Aaronson and Vazirani [10, 11] and, in fact, Scott Aaronson has offered a 25$ prize for its resolution [10]. While, as of yet, the question remains open, one does arrive at a positive answer through slight alterations of the interactive-proof system. Specifically, if the verifier interacts with two or more BQP-restricted provers, instead of one, and the provers are not allowed to communicate with each other during the protocol, then it is possible to efficiently verify arbitrary BQP computations [19–25]. Alternatively, in the single-prover setting, if we allow the verifier to have a constant-size quantum computer and the ability to send/receive quantum states to/from the prover then it is again possible to verify all polynomial-time quantum computations [26–34]. Note that in this case, while the verifier is no longer fully "classical", its computational capability is still restricted to BPP since simulating a constant-size quantum computer can be done in constant time. These scenarios are depicted in Figure 2.

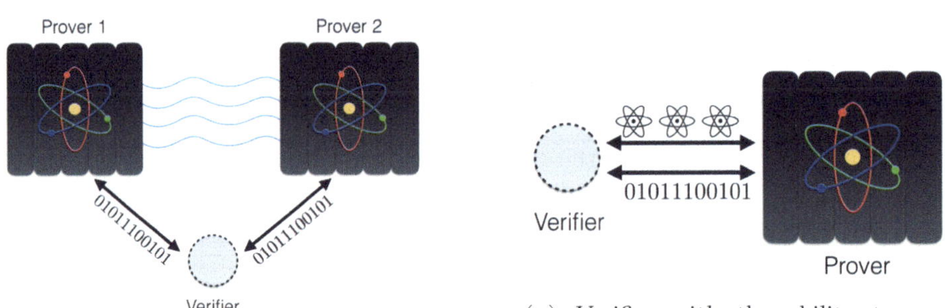

(A) Classical verifier interacting with two entangled but non-communicating quantum provers

(B) Verifier with the ability to prepare or measure constant-size quantum states interacting with a single quantum prover

FIGURE 2. Models for verifiable quantum computation

The primary technique that has been employed in most, thought not all, of these settings, to achieve verification, is known as *blindness*. This entails delegating a computation to the provers in such a way that they cannot distinguish this computation from any other of the same size, unconditionally[4]. Intuitively, verification then follows by having most of these computations be *tests* or *traps* which the verifier can check. If the provers attempt to deviate they will have a high chance of triggering these traps and prompt the verifier to reject.

In this paper, we review all of these approaches to verification. We broadly classify the protocols as follows:

1. **Single-prover prepare-and-send.** These are protocols in which the verifier has the ability to prepare quantum states and send them to the prover. They are covered in Section 2.
2. **Single-prover receive-and-measure.** In this case, the verifier receives quantum states from the prover and has the ability to measure them. These protocols are presented in Section 3.
3. **Multi-prover entanglement-based.** In this case, the verifier is fully classical, however it interacts with more than one prover. The provers are not allowed to communicate during the protocol. Section 4 is devoted to these protocols.

From the complexity-theoretic perspective, the protocols from the first two sections are classified as QPIP protocols, or protocols in which the verifier has a minimal quantum device and can send or receive quantum states. Conversely, the entanglement-based protocols are classified as MIP* protocols, in which the verifier is classical and interacting with provers that share entanglement[5].

After reviewing the major approaches to verification, in Section 5, we address a number of related topics. In particular, while all of the protocols from Sections 2-4 are concerned with the verification of general BQP computations, in Subsection 5.1 we mention *sub-universal* protocols, designed to verify only a particular subclass of quantum computations. Next, in Subsection 5.2 we discuss an important practical aspect concerning verification, which is *fault tolerance*. We comment on the possibility of making protocols resistant to noise which could affect any of the involved quantum devices. This is an important consideration for any realistic implementation of a verification protocol. Finally, in Subsection 5.3 we outline some of the existing experimental implementations of these protocols.

Throughout the review, we are assuming familiarity with the basics of quantum information theory and some elements of complexity theory. However, we provide a brief overview of these topics as well as other notions that are used in this review (such as *measurement-based quantum computing*) in the appendix, Section 7. Note also, that we will be referencing complexity classes such as BQP,

[4]In other words, the provers would not be able to differentiate among the different computations even if they had unbounded computational power.

[5]Strictly speaking, QPIP and MIP* are classes of decision problems. Hence, whenever we say, for instance, "protocol X is a QPIP (MIP*) protocol" we mean that each problem in the class QPIP (MIP*) can be decided by protocol X. When the provers are restricted to polynomial-time quantum computations, both of these classes are equal to BQP. However, the reason for specifying either QPIP or MIP* is to emphasize the structure of the protocol (i.e., whether it utilizes a single prover and a minimally quantum verifier or a classical verifier and multiple entangled provers) as it follows from the definitions of these classes. The definitions can be found in Subsection 7.3.

QMA, QPIP and MIP*. Definitions for all of these are provided in Subsection 7.3 of the appendix. We begin with a short overview of *blind quantum computing.*

1.1. Blind Quantum Computing

The concept of blind computing is highly relevant to quantum verification. Here, we simply give a succinct outline of the subject. For more details, see this review of blind quantum computing protocols by Fitzsimons [35] as well as [36–40]. Note that, while the review of Fitzsimons covers all of the material presented in this section (and more), we restate the main ideas, so that our review is self-consistent and also in order to establish some of the notation that is used throughout the rest of the paper.

Blindness is related to the idea of *computing on encrypted data* [41]. Suppose a client has some input x and would like to compute a function f of that input, however, evaluating the function directly is computationally infeasible for the client. Luckily, the client has access to a server with the ability to evaluate $f(x)$. The problem is that the client does not trust the server with the input x, since it might involve private or secret information (e.g., medical records, military secrets, proprietary information etc). The client does, however, have the ability to encrypt x, using some encryption procedure $\mathcal{E}$, to a ciphertext $y \leftarrow \mathcal{E}(x)$. As long as this encryption procedure hides x sufficiently well, the client can send y to the server and receive in return (potentially after some interaction with the server) a string z which decrypts to $f(x)$. In other words, $f(x) \leftarrow \mathcal{D}(z)$, where $\mathcal{D}$ is a decryption procedure that can be performed efficiently by the client[6]. The encryption procedure can, roughly, provide two types of security: *computational* or *information-theoretic.* Computational security means that the protocol is secure as long as certain computational assumptions are true (for instance that the server is unable to invert one-way functions). Information-theoretic security (sometimes referred to as unconditional security), on the other hand, guarantees that the protocol is secure even against a server of unbounded computational power. See [46] for more details on these topics.

In the quantum setting, the situation is similar to that of QPIP protocols: the client is restricted to BPP computations, but has some limited quantum capabilities, whereas the server is a BQP machine. Thus, the client would like to delegate BQP functions to the server, while keeping the input and the output hidden. The first solution to this problem was provided by Childs [36]. His protocol achieves information-theoretic security but also requires the client and the server to exchange quantum messages for a number of rounds that is proportional to the size

[6]In the classical setting, computing on encrypted data culminated with the development of *fully homomorphic encryption* (FHE), which is considered the "*holly grail*" of the field [42–45]. Using FHE, a client can delegate the evaluation of *any* polynomial-size classical circuit to a server, such that the input and output of the circuit are kept hidden from the server, based on reasonable computational assumptions. Moreover, the protocol involves only one round of back-and-forth interaction between client and server.

of the computation. This was later improved in a protocol by Broadbent, Fitzsimons and Kashefi [37], known as *universal blind quantum computing* (UBQC), which maintained information-theoretic security but reduced the quantum communication to a single message from the client to the server. UBQC still requires the client and the server to have a total communication which is proportional to the size of the computation, however, apart from the first quantum message, the interaction is purely classical. Let us now state the definition of perfect, or information-theoretic, blindness from [37]:

Definition 1 (Blindness). *Let* P *be a delegated quantum computation protocol involving a client and a server. The client draws the input from the random variable* X*. Let* $L(X)$ *be any function of this random variable. We say that the protocol is* blind while leaking at most L(X) *if, on the client's input* X*, for any* $l \in Range(L)$*, the following two hold when given* $l \leftarrow L(X)$*:*

1. *The distribution of the classical information obtained by the server in* P *is independent of* X*.*
2. *Given the distribution of classical information described in* 1*, the state of the quantum system obtained by the server in* P *is fixed and independent of* X*.*

The definition is essentially saying that the server's "view" of the protocol should be independent of the input, when given the length of the input. This view consists, on the one hand, of the classical information he receives, which is independent of X, given $L(X)$. On the other hand, for any fixed choice of this classical information, his quantum state should also be independent of X, given $L(X)$. Note that the definition can be extended to the case of multiple servers as well. To provide intuition for how a protocol can achieve blindness, we will briefly recap the main ideas from [36, 37]. We start by considering the *quantum one-time pad.*

1.1.1. Quantum one-time pad. Suppose we have two parties, Alice and Bob, and Alice wishes to send one qubit, ρ, to Bob such that all information about ρ is kept hidden from a potential eavesdropper, Eve. For this to work, we will assume that Alice and Bob share two classical random bits, denoted b_1 and b_2, that are known only to them. Alice will then apply the operation $\mathsf{X}^{b_1}\mathsf{Z}^{b_2}$ (the quantum one-time pad) to ρ, resulting in the state $\mathsf{X}^{b_1}\mathsf{Z}^{b_2}\rho\mathsf{Z}^{b_2}\mathsf{X}^{b_1}$, and send this state to Bob. If Bob then also applies $\mathsf{X}^{b_1}\mathsf{Z}^{b_2}$ to the state he received, he will recover ρ. What happens if Eve intercepts the state that Alice sends to Bob? Because Eve does not know the random bits b_1 and b_2, the state that she will intercept will be:

$$\frac{1}{4}\sum_{b_1,b_2\in\{0,1\}} \mathsf{X}^{b_1}\mathsf{Z}^{b_2}\rho\mathsf{Z}^{b_2}\mathsf{X}^{b_1}. \tag{1}$$

However, it can be shown that for any single-qubit state ρ:

$$\frac{1}{4}\sum_{b_1,b_2\in\{0,1\}} \mathsf{X}^{b_1}\mathsf{Z}^{b_2}\rho\mathsf{Z}^{b_2}\mathsf{X}^{b_1} = I/2\,. \tag{2}$$

In other words, the state that Eve intercepts is the totally mixed state, *irrespective of the original state* ρ. But the totally mixed state is, by definition, the state of maximal uncertainty. Hence, Eve cannot recover any information about ρ, regardless of her computational power. Note, that for this argument to work, and in particular for Equation 2 to be true, Alice and Bob's shared bits must be *uniformly random*. If Alice wishes to send n qubits to Bob, then as long as Alice and Bob share $2n$ random bits, they can simply perform the same procedure for each of the n qubits. Equation 2 generalizes for the multi-qubit case so that for an n-qubit state ρ we have:

$$\frac{1}{4^n} \sum_{\mathbf{b_1}, \mathbf{b_2} \in \{0,1\}^n} \mathsf{X}(\mathbf{b_1})\mathsf{Z}(\mathbf{b_2})\rho\mathsf{Z}(\mathbf{b_2})\mathsf{X}(\mathbf{b_1}) = I/2^n . \tag{3}$$

Here, $\mathbf{b_1}$ and $\mathbf{b_2}$ are n-bit vectors, $\mathsf{X}(\mathbf{b}) = \bigotimes_{i=1}^{n} \mathsf{X}^{\mathbf{b(i)}}$, $\mathsf{Z}(\mathbf{b}) = \bigotimes_{i=1}^{n} \mathsf{Z}^{\mathbf{b(i)}}$ and I is the 2^n-dimensional identity matrix.

1.1.2. Childs' protocol for blind computation. Now suppose Alice has some n-qubit state ρ and wants a quantum circuit $\mathcal{C}$ to be applied to this state and the output to be measured in the computational basis. However, she only has the ability to store n qubits, prepare qubits in the $|0\rangle$ state, swap any two qubits, or apply a Pauli X or Z to any of the n qubits. So in general, she will not be able to apply a general quantum circuit $\mathcal{C}$, or perform measurements. Bob, on the other hand, does not have these limitations as he is a BQP machine and thus able to perform universal quantum computations. How can Alice delegate the application of $\mathcal{C}$ to her state without revealing any information about it, apart from its size, to Bob? The answer is provided by Childs' protocol [36]. Before presenting the protocol, recall that any quantum circuit, $\mathcal{C}$, can be expressed as a combination of Clifford operations and T gates. Additionally, Clifford operations commute with Pauli gates. All of these notions are defined in the appendix, Subsection 7.1.6.

First, Alice will one-time pad her state and send the padded state to Bob. As mentioned, this will reveal no information to Bob about ρ. Next, Alice instructs Bob to start applying the gates in $\mathcal{C}$ to the padded state. Apart from the T gates, all other operations in $\mathcal{C}$ will be Clifford operations, which commute with Pauli gates. Thus, if Alice's padded state is $\mathsf{X}(\mathbf{b_1})\mathsf{Z}(\mathbf{b_2})\rho\mathsf{Z}(\mathbf{b_2})\mathsf{X}(\mathbf{b_1})$ and Bob applies the Clifford unitary U_C, the resulting state will be:

$$U_C\mathsf{X}(\mathbf{b_1})\mathsf{Z}(\mathbf{b_2})\rho\mathsf{Z}(\mathbf{b_2})\mathsf{X}(\mathbf{b_1})U_C^\dagger = \mathsf{X}(\mathbf{b_1'})\mathsf{Z}(\mathbf{b_2'})U_C\rho U_C^\dagger\mathsf{Z}(\mathbf{b_2'})\mathsf{X}(\mathbf{b_1'}) . \tag{4}$$

Here, $\mathbf{b_1'}$ and $\mathbf{b_2'}$ are linearly related to $\mathbf{b_1}$ and $\mathbf{b_2}$, meaning that Alice can compute them using only *xor* operations. This gives her an updated pad for her state. If $\mathcal{C}$ consisted exclusively of Clifford operations then Alice would only need to keep track of the updated pad (also referred to as the *Pauli frame*) after each gate. Once Bob returns the state, she simply undoes the one-time pad using the updated key, that she computed, and recovers $\mathcal{C}\rho\mathcal{C}^\dagger$. Of course, this will not work if $\mathcal{C}$ contains

T gates, since:

$$\mathsf{TX}^a = \mathsf{X}^a\mathsf{S}^a\mathsf{T}, \tag{5}$$

where $\mathsf{S} = \mathsf{T}^2$. In other words, if we try to commute the T operation with the one-time pad we will get an unwanted S gate applied to the state. Worse, the S will have a dependency on one of the secret pad bits for that particular qubit. This means that if Alice asks Bob to apply an S^a operation she will reveal one of her pad bits. Fortunately, as explained in [36], there is a simple way to remedy this problem. After each T gate, Alice asks Bob to return the quantum state to her. Suppose that Bob had to apply a T on qubit j. Alice then applies a new one-time pad on that qubit. If the previous pad had no X gate applied to j, she will swap this qubit with a dummy state that does not take part in the computation[7], otherwise she leaves the state unchanged. She then returns the state to Bob and asks him to apply an S gate to qubit j. Since this operation will always be applied, after a T gate, it does not reveal any information about Alice's pad. Bob's operation will therefore cancel the unwanted S gate when this appears and otherwise it will act on a qubit which does not take part in the computation. The state should then be sent back to Alice so that she can undo the swap operation if it was performed. Once all the gates in $\mathcal{C}$ have been applied, Bob is instructed to measure the resulting state in the computational basis and return the classical outcomes to Alice. Since the quantum output was one-time padded, the classical outcomes will also be one-time padded. Alice will then undo the pad an recover her desired output.

While Childs' protocol provides an elegant solution to the problem of quantum computing on encrypted data, it has significant requirements in terms of Alice's quantum capabilities. If Alice's input is fully classical, i.e., some state $|x\rangle$, where $x \in \{0,1\}^n$, then Alice would only require a constant-size quantum memory. Even so, the protocol requires Alice and Bob to exchange multiple quantum messages. This, however, is not the case with UBQC which limits the quantum communication to one quantum message sent from Alice to Bob at the beginning of the protocol. Let us now briefly state the main ideas of that protocol.

1.1.3. Universal Blind Quantum Computation (UBQC). In UBQC the objective is to not only hide the input (and output) from Bob, but also the circuit which will act on that input[8] [37]. As in the previous case, Alice would like to delegate to Bob the application of some circuit $\mathcal{C}$ on her input (which, for simplicity, we will assume is classical). This time, however, we view $\mathcal{C}$ as an MBQC computation[9]. By considering some universal graph state, $|G\rangle$, such as the brickwork state (see Figure 17), Alice can convert $\mathcal{C}$ into a description of $|G\rangle$ (the graph G) along with the appropriate measurement angles for the qubits in the graph state. By the

[7]For instance, her initial state ρ could contain a number of $|0\rangle$ qubits that is equal to the number of T gates in the circuit.

[8]This is also possible in Childs' protocol by simply encoding the description of the circuit $\mathcal{C}$ in the input and asking Bob to run a universal quantum circuit. The one-time padded input that is sent to Bob would then comprise of both the description of $\mathcal{C}$ as well as x, the input for $\mathcal{C}$.

[9]For a brief overview of MBQC see Subsection 7.2.

property of the universal graph state, the graph G would be the same for all circuits $\mathcal{C}'$ having the same number of gates as $\mathcal{C}$. Hence, if she were to send this description to Bob, it would not reveal to him the circuit $\mathcal{C}$, merely an upper bound on its size. It is, in fact, the measurement angles and the ordering of the measurements (known as *flow*) that uniquely characterise $\mathcal{C}$ [47]. But the measurement angles are chosen assuming all qubits in the graph state were initially prepared in the $|+\rangle$ state. Since these are XY-plane measurements, as explained in Subsection 7.1, the probabilities, for the two possible outcomes, depend only on the difference between the measurement angle and the preparation angle of the state, which is 0, in this case[10]. Suppose instead that each qubit, indexed i, in the cluster state, were instead prepared in the state $|+_{\theta_i}\rangle$. Then, if the original measurement angle for qubit i was ϕ_i, to preserve the relative angles, the new value would be $\phi_i + \theta_i$. If the values for θ_i are chosen at random, then they effectively act as a one-time pad for the original measurement angles ϕ_i. This means that if Bob does not know the preparation angles of the qubits and were instructed to measure them at the updated angles $\phi_i + \theta_i$, to him, these angles would be indistinguishable from random, irrespective of the values of ϕ_i. He would, however, learn the measurement outcomes of the MBQC computation. But there is a simple way to hide this information as well. One can flip the probabilities of the measurement outcomes for a particular state by performing a π rotation around Z axis. In other words, the updated measurement angles will be $\delta_i = \phi_i + \theta_i + r_i\pi$, where r_i is sampled randomly from $\{0, 1\}$.

To recap, UBQC works as follows:

(1) Alice chooses an input x and a quantum computation $\mathcal{C}$ that she would like Bob to perform on $|x\rangle$.

(2) She converts x and $\mathcal{C}$ into a pair $(G, \{\phi_i\}_i)$, where $|G\rangle$ is an N-qubit universal graph state (with an established ordering for measuring the qubits), $N = O(|\mathcal{C}|)$ and $\{\phi_i\}_i$ is the set of computation angles allowing for the MBQC computation of $\mathcal{C}\,|x\rangle$.

(3) She picks, uniformly at random, values θ_i, with i going from 1 to N, from the set $\{0, \pi/4, 2\pi/4, \ldots, 7\pi/4\}$ as well as values r_i from the set $\{0, 1\}$.

(4) She then prepares the states $|+_{\theta_i}\rangle$ and sends them to Bob, who is instructed to entangle them, using CZ operations, according to the graph structure G.

(5) Alice then asks Bob to measure the qubits at the angles $\delta_i = \phi'_i + \theta_i + r_i\pi$ and return the measurement outcomes to her. Here, ϕ'_i is an updated version of ϕ_i that incorporates corrections resulting from previous measurements, as in the description of MBQC given in Subsection 7.2.

(6) After all the measurements have been performed, Alice undoes the r_i one-time padding of the measurement outcomes, thus recovering the true outcome of the computation.

[10]This remains true even if the qubits have been entangled with the CZ operation.

The protocol is illustrated schematically in Figure 3, reproduced from [48] (the variables b_1, b_2, b_3 indicate measurement outcomes).

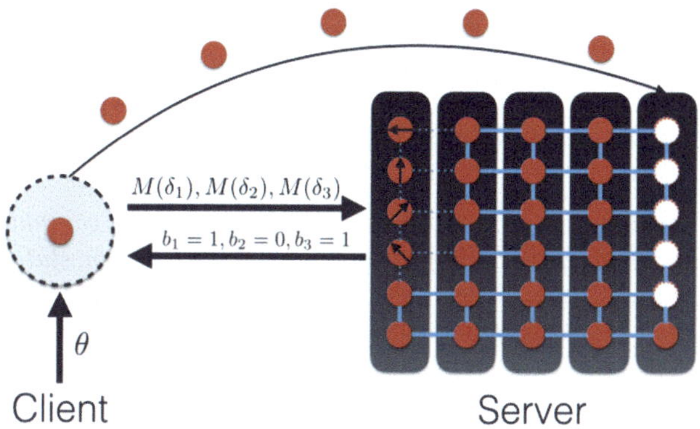

FIGURE 3. Universal Blind Quantum Computation

We can see that as long as Bob does not know the values of the θ_i and r_i variables, the measurements he is asked to perform, as well as their outcomes, will appear totally random to him. The reason why Bob cannot learn the values of θ_i and r_i from the qubits prepared by Alice is due to the limitation, in quantum mechanics, that one cannot distinguish between non-orthogonal states. In fact, a subsequent paper by Dunjko and Kashefi shows that Alice can utilize any two non-overlapping, non-orthogonal states in order to perform UBQC [49].

2. Prepare-and-Send Protocols

We start by reviewing QPIP protocols in which the only quantum capability of the verifier is to prepare and send constant-size quantum states to the prover (no measurement). The verifier must use this capability in order to delegate the application of some BQP circuit, $\mathcal{C}$, on an input $|\psi\rangle$[11]. Through interaction with the prover, the verifier will attempt to certify that the correct circuit was indeed applied on her input, with high probability, aborting the protocol otherwise.

There are three major approaches that fit this description and we devote a subsection to each of them:

1. **Subsection 2.1**: two protocols based on quantum authentication, developed by Aharonov, Ben-Or, Eban and Mahadev [26, 27].
2. **Subsection 2.2**: a trap-based protocol, developed by Fitzsimons and Kashefi [28].
3. **Subsection 2.3**: a scheme based on repeating indistinguishable runs of tests and computations, developed by Broadbent [29].

[11]This input can be a classical bit string $|x\rangle$, though it can also be more general.

In the context of prepare-and-send protocols, it is useful to provide more refined notions of completeness and soundness than the ones in the definition of a QPIP protocol. This is because, apart from knowing that the verifier wishes to delegate a BQP computation to the prover, we also know that it prepares a particular quantum state and sends it to the prover to act with some unitary operation on it (corresponding to the quantum circuit associated with the BQP computation). This extra information allows us to define *δ-correctness* and *ϵ-verifiability.* We start with the latter:

Definition 2 (ϵ-verifiability). *Consider a delegated quantum computation protocol between a verifier and a prover and let the verifier's quantum state be $|\psi\rangle\,|flag\rangle$, where $|\psi\rangle$ is the input state to the protocol and $|flag\rangle$ is a flag state denoting whether the verifier accepts ($|flag\rangle = |acc\rangle$) or rejects ($|flag\rangle = |rej\rangle$) at the end of the protocol. Consider also the quantum channel Enc_s (encoding), acting on the verifier's state, where s denotes a private random string, sampled by the verifier from some distribution $p(s)$. Let $\mathcal{P}_{honest}$ denote the CPTP map corresponding to the honest action of the prover in the protocol (i.e., following the instructions of the verifier) acting on the verifier's state. Additionally, define:*

$$P^s_{incorrect} = (I - |\Psi^s_{out}\rangle\langle\Psi^s_{out}|) \otimes |acc^s\rangle\langle acc^s| \tag{6}$$

as a projection onto the orthogonal complement of the correct output:

$$|\Psi^s_{out}\rangle\langle\Psi^s_{out}| = Tr_{flag}(\mathcal{P}_{honest}(Enc_s(|\psi\rangle\langle\psi| \otimes |acc\rangle\langle acc|))) \tag{7}$$

and on acceptance for the flag state:

$$|acc^s\rangle\langle acc^s| = Tr_{input}(\mathcal{P}_{honest}(Enc_s(|\psi\rangle\langle\psi| \otimes |acc\rangle\langle acc|))). \tag{8}$$

We say that such a protocol is ϵ-verifiable (with $0 \leq \epsilon \leq 1$), if for any action $\mathcal{P}$, of the prover, we have that[12]*:*

$$Tr\left(\sum_s p(s) P^s_{incorrect} \mathcal{P}(Enc_s(|\psi\rangle\langle\psi| \otimes |acc\rangle\langle acc|))\right) \leq \epsilon. \tag{9}$$

Essentially, this definition says that the probability for the output of the protocol to be incorrect *and* the verifier accepting, should be bounded by ϵ. As a simple mathematical statement we would write this as the joint distribution:

$$Pr(incorrect, accept) \leq \epsilon. \tag{10}$$

One could also ask whether $Pr(incorrect|accept)$ should also be upper bounded. Indeed, it would seem like this conditional distribution is a better match for our intuition regarding the "*probability of accepting an incorrect outcome*". However, upon closer scrutiny, we find that the conditional distribution cannot be upper

---

[12]An alternative to Equation 9 is: $TD(\rho_{out}, p|\Psi^s_{out}\rangle\langle\Psi^s_{out}| \otimes |acc^s\rangle\langle acc^s| + (1-p)\rho \otimes |rej^s\rangle\langle rej^s|) \leq \epsilon$, for some $0 \leq p \leq 1$ and some density matrix ρ. In other words, the output state of the protocol, ρ_{out}, is close to a state which is a mixture of the correct output state with acceptance and an arbitrary state and rejection. This definition can be more useful when one is interested in a *quantum output* for the protocol (i.e., the prover returns a quantum state to the verifier). Such a situation is particularly useful when composing verification protocols [20,50–52].

bounded. To understand why, note that we can express the conditional distribution as:

$$Pr(incorrect|accept) = \frac{Pr(incorrect, accept)}{Pr(accept)}. \tag{11}$$

Now, it is true that if $Pr(accept)$ is close to 1 *and* the joint distribution is upper bounded, then the conditional distribution will also be upper bounded. However, suppose the prover behaves in such a way that it causes the verifier to reject most of the time. Moreover, suppose that the joint distribution $Pr(incorrect, accept)$ has a fixed but non-zero value[13]. In that case, the probability of acceptance would be very small and we can no longer bound the conditional probability. Therefore, the quantity that we will generally be interested in, and the one that will always be bounded in verification protocols is the joint probability $Pr(incorrect, accept)$.

We now define δ-correctness:

Definition 3 (δ-correctness). *Consider a delegated quantum computation protocol between a verifier and a prover. Using the notation from Definition 2, and letting:*

$$P^s_{correct} = |\Psi^s_{out}\rangle\langle\Psi^s_{out}| \otimes |acc^s\rangle\langle acc^s| \tag{12}$$

be the projection onto the correct output and on acceptance for the flag state, we say that such a protocol is δ-correct (with $0 \leq \delta \leq 1$), if for all strings s we have that:

$$Tr\left(P^s_{correct}\mathcal{P}_{honest}(Enc_s(|\psi\rangle\langle\psi| \otimes |acc\rangle\langle acc|))\right) \geq \delta\,. \tag{13}$$

This definition says that when the prover behaves honestly, the verifier obtains the correct outcome, with high probability, for any possible choice of its secret parameters.

If a prepare-and-send protocol has both δ-correctness and ϵ-verifiability, for some $\delta > 0$, $\epsilon < 1$, it will also have completeness $\delta(1/2+1/poly(n))$ and soundness ϵ as a QPIP protocol, where n is the size of the input. The reason for the asymmetry in completeness and soundness is that in the definition of δ-correctness we require that the output quantum state of the protocol is δ-close to the output quantum state of the desired computation. But the computation outcome is dictated by a measurement of this state, which succeeds with probability at least $1/2+1/poly(n)$, from the definition of BQP. Combining these facts leads to $\delta(1/2 + 1/poly(n))$ completeness. It follows that for this to be a valid QPIP protocol it must be that $\delta(1/2 + 1/poly(n)) - \epsilon \geq 1/poly(n)$, for all inputs. For simplicity, we will instead require $\delta/2 - \epsilon \geq 1/poly(n)$, which implies the previous inequality. As we will see, for all prepare-and-send protocols $\delta = 1$. This condition is easy to achieve by simply designing the protocol so that the honest behaviour of the prover leads to the correct unitary being applied to the verifier's quantum state. Therefore, the main challenge with these protocols will be to show that $\epsilon \leq 1/2 - 1/poly(n)$.

---

[13]For instance, if the prover provides random responses to the verifier there is a non-zero chance that one instance of these responses will pass all of the verifier's tests and cause it to accept an incorrect outcome.

2.1. Quantum Authentication-Based Verification

This subsection is dedicated to the two protocols presented in [26,27] by Aharonov et al. These protocols are extensions of *Quantum Authentication Schemes* (QAS), a security primitive introduced in [53] by Barnum et al. A QAS is a scheme for transmitting a quantum state over an insecure quantum channel and being able to indicate whether the state was corrupted or not. More precisely, a QAS involves a *sender* and a *receiver*. The sender has some quantum state $|\psi\rangle |flag\rangle$ that it would like to send to the receiver over an insecure channel. The state $|\psi\rangle$ is the one to be authenticated, while $|flag\rangle$ is an indicator state used to check whether the authentication was performed successfully. We will assume that $|flag\rangle$ starts in the state $|acc\rangle$. It is also assumed that the sender and the receiver share some classical key k, drawn from a probability distribution $p(k)$. To be able to detect the effects of the insecure channel on the state, the sender will first apply some encoding procedure Enc_k thus obtaining $\rho = \sum_k p(k) Enc_k(|\psi\rangle |acc\rangle)$. This state is then sent over the quantum channel where it can be tampered with by an eavesdropper resulting in a new state ρ'. The receiver, will then apply a decoding procedure to this state, resulting in $Dec_k(\rho')$ and decide whether to accept or reject by measuring the flag subsystem[14]. Similar to verification, this protocol must satisfy two properties:

1. **δ-correctness**. Intuitively this says that if the state sent through the channel was not tampered with, then the receiver should accept with high probability (at least δ), irrespective of the used keys. More formally, for $0 \leq \delta \leq 1$, let:
$$P_{correct} = |\psi\rangle \langle\psi| \otimes |acc\rangle \langle acc|$$
be the projector onto the correct state $|\psi\rangle$ and on acceptance for the flag state. Then, it must be the case that for all keys k:
$$Tr\left(P_{correct} Dec_k(Enc_k(|\psi\rangle \langle\psi| \otimes |acc\rangle \langle acc|))\right) \geq \delta$$
2. **ϵ-security**. This property states that for any deviation that the eavesdropper applies on the sent state, the probability that the resulting state is far from ideal *and* the receiver accepts is small. Formally, for $0 \leq \epsilon \leq 1$, let:
$$P_{incorrect} = (I - |\psi\rangle \langle\psi|) \otimes |acc\rangle \langle acc|$$
be the projector onto the orthogonal complement of the correct state $|\psi\rangle$, and on acceptance, for the flag state. Then, it must be the case that for any CPTP action, $\mathcal{E}$, of the eavesdropper, we have:
$$Tr\left(P_{incorrect} \sum_k p(k) Dec_k(\mathcal{E}(Enc_k(|\psi\rangle \langle\psi| \otimes |acc\rangle \langle acc|)))\right) \leq \epsilon\,.$$

To make the similarities between QAS and prepare-and-send protocols more explicit, suppose that, in the above scheme, the receiver were trying to authenticate the state $U |\psi\rangle$ instead of $|\psi\rangle$, for some unitary U. In that case, we could view

[14] The projectors for the measurement are assumed to be $P_{acc} = |acc\rangle \langle acc|$, for acceptance and $P_{rej} = I - |acc\rangle \langle acc|$ for rejection.

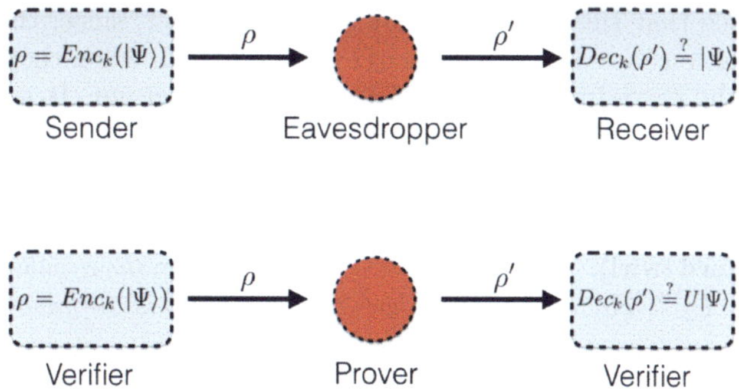

FIGURE 4. QAS-based verification

the sender as the verifier at the beginning of the protocol, the eavesdropper as the prover and the receiver as the verifier at the end of the protocol. This is illustrated in Figure 4, reproduced from [48]. If one could therefore augment a QAS scheme with the ability of applying a quantum circuit on the state, while keeping it authenticated, then one would essentially have a prepare-and-send verification protocol. This is what is achieved by the two protocols of Aharonov et al.

2.1.1. Clifford-QAS VQC. The first protocol, named *Clifford QAS-based Verifiable Quantum Computing* (Clifford-QAS VQC) is based on a QAS which uses Clifford operations in order to perform the encoding procedure. Strictly speaking, this protocol is not a prepare-and-send protocol, since, as we will see, it involves the verifier performing measurements as well. However, it is a precursor to the second protocol from [26,27], which is a prepare-and-send protocol. Hence, why we review the Clifford-QAS VQC protocol here.

Let us start by explaining the authentication scheme first. As before, let $|\psi\rangle |flag\rangle$ be the state that the sender wishes to send to the receiver and k be their shared random key. We will assume that $|\psi\rangle$ is an n-qubit state, while $|flag\rangle$ is an m-qubit state. Let $t = n + m$ and $\mathfrak{C}_t$ be the set of t-qubit Clifford operations. We also assume that each possible key, k, can specify a unique t-qubit Clifford operation, denoted C_k[15]. The QAS works as follows:

(1) The sender performs the encoding procedure Enc_k. This consists of applying the Clifford operation C_k to the state $|\psi\rangle |acc\rangle$.

(2) The state is sent through the quantum channel.

(3) The receiver applies the decoding procedure Dec_k which consists of applying $C_k^\dagger$ to the received state.

(4) The receiver measures the $flag$ subsystem and accepts if it is in the $|acc\rangle$ state.

[15]Hence $|k| = O(\log(|\mathfrak{C}_t|))$.

We can see that this protocol has correctness $\delta = 1$, since, the sender and receiver's operations are exact inverses of each other and, when there is no intervention from the eavesdropper, they will perfectly cancel out. It is also not too difficult to show that the protocol achieves security $\epsilon = 2^{-m}$. We will include a sketch proof of this result as all other proofs of security, for prepare-and-send protocols, rely on similar ideas. Aharonov et al start by using the following lemma:

Lemma 1 (Clifford twirl). *Let P_1, P_2 be two operators from the n-qubit Pauli group, such that $P_1 \neq P_2$[16]. For any n-qubit density matrix ρ it is the case that:*

$$\sum_{C \in \mathfrak{C}_n} C^\dagger P_1 C \rho C^\dagger P_2 C = 0\,. \tag{14}$$

To see how this lemma is applied, recall that any CPTP map admits a Kraus decomposition, so we can express the eavesdropper's action as:

$$\mathcal{E}(\rho) = \sum_i K_i \rho K_i^\dagger, \tag{15}$$

where, $\{K_i\}_i$ is the set of Kraus operators, satisfying:

$$\sum_i K_i K_i^\dagger = I\,. \tag{16}$$

Additionally, recall that the n-qubit Pauli group is a basis for all $2^n \times 2^n$ matrices, which means that we can express each Kraus operator as:

$$K_i = \sum_j \alpha_{ij} P_j\,, \tag{17}$$

where j ranges over all indices for n-qubit Pauli operators and $\{\alpha_{ij}\}_{i,j}$ is a set of complex numbers such that:

$$\sum_{i,j} \alpha_{ij} \alpha_{ij}^* = 1\,. \tag{18}$$

For simplicity, assume that the phase information of each Pauli operator, i.e., whether it is $+1$, -1, $+i$ or $-i$, is absorbed in the α_{ij} terms. One can then re-express the eavesdropper's deviation as:

$$\mathcal{E}(\rho) = \sum_{ijk} \alpha_{ij} \alpha_{ik}^* P_j \rho P_k\,. \tag{19}$$

We would now like to use Lemma 1 to see how this deviation affects the encoded state. Given that the encoding procedure involves applying a random Clifford operation to the initial state, which we will denote $|\Psi_{in}\rangle = |\psi\rangle\,|acc\rangle$, the state received by the eavesdropper will be:

$$\rho = \frac{1}{|\mathfrak{C}_t|} \sum_l C_l \,|\Psi_{in}\rangle\,\langle\Psi_{in}|\, C_l^\dagger\,. \tag{20}$$

[16] Technically, what is required here is that $|P_1| \neq |P_2|$, since global phases are ignored.

Acting with $\mathcal{E}$ on this state and using Equation 19 yields:

$$\mathcal{E}(\rho) = \frac{1}{|\mathfrak{C}_t|} \sum_{ijkl} \alpha_{ij}\alpha^*_{ik} P_j C_l \left|\Psi_{in}\right\rangle \left\langle\Psi_{in}\right| C_l^\dagger P_k \,. \tag{21}$$

Now using Lemma 1 we can see that all terms which act with different Pauli operations on both sides of $C_l \left|\Psi_{in}\right\rangle \left\langle\Psi_{in}\right| C_l^\dagger$ will vanish, resulting in:

$$\mathcal{E}(\rho) = \frac{1}{|\mathfrak{C}_t|} \sum_{ijl} \alpha_{ij}\alpha^*_{ij} P_j C_l \left|\Psi_{in}\right\rangle \left\langle\Psi_{in}\right| C_l^\dagger P_j \,, \tag{22}$$

which is a convex combination of Pauli operations acting on the encoded state[17]. The receiver takes this state and applies the decoding operation, which involves inverting the Clifford that was applied by the sender. This will produce the state:

$$\sigma = \frac{1}{|\mathfrak{C}_t|} \sum_{ijl} \alpha_{ij}\alpha^*_{ij} C_l^\dagger P_j C_l \left|\Psi_{in}\right\rangle \left\langle\Psi_{in}\right| C_l^\dagger P_j C_l \,. \tag{23}$$

Let us take a step back and understand what happened. We saw that any general map can be expressed as a combination of Pauli operators acting on both sides of the target state, ρ. Importantly, the Pauli operators on both sides needed not be equal. However, if the target state is an equal mixture of Clifford operations applied on some other state, in our case $\left|\Psi_{in}\right\rangle \left\langle\Psi_{in}\right|$, the Clifford twirl lemma makes all non-equal Pauli terms vanish. One is then left with a convex combination of Pauli operators on the target state. If one then undoes the random Clifford operation on this state, each Pauli term in the convex combination will be conjugated by all Cliffords in the set $\mathfrak{C}_t$. Since conjugating a Pauli by Cliffords results in a new Pauli, it can be shown that the state becomes a mixture of the original $\left|\Psi_{in}\right\rangle \left\langle\Psi_{in}\right|$ and a *uniform* convex combination of non-identity Paulis acting on $\left|\Psi_{in}\right\rangle \left\langle\Psi_{in}\right|$. Mathematically, this means:

$$\sigma = \beta \left|\Psi_{in}\right\rangle \left\langle\Psi_{in}\right| + \frac{1-\beta}{4^t - 1} \sum_{i, P_i \neq I} P_i \left|\Psi_{in}\right\rangle \left\langle\Psi_{in}\right| P_i \tag{24}$$

where $0 \leq \beta \leq 1$.

The last element in the proof is to compute $Tr(P_{incorrect}\sigma)$. Since the first term in the mixture is the ideal state, we will be left with:

$$Tr(P_{incorrect}\sigma) = \frac{1-\beta}{4^t - 1} \sum_{i, P_i \neq I} Tr(P_{incorrect} P_i \left|\Psi_{in}\right\rangle \left\langle\Psi_{in}\right| P_i) \,. \tag{25}$$

The terms in the summation will be non-zero whenever P_i acts as identity on the flag subsystem. The number of such terms can be computed to be exactly $4^n 2^m - 1$ and using the fact that $t = m + n$ and $1 - \beta \leq 1$, we have:

$$Tr(P_{incorrect}\sigma) \leq (1-\beta)\frac{4^n 2^m - 1}{4^{m+n}} \leq \frac{1}{2^m} \tag{26}$$

concluding the proof.

[17]This is because $\alpha_{ij}\alpha^*_{ij} = |\alpha_{ij}|^2$ is a positive real number and $\sum_{i,j} \alpha_{ij}\alpha^*_{ij} = 1$.

As mentioned, in all prepare-and-send protocols we assume that the verifier will prepare some state $|\psi\rangle$ on which it wants to apply a quantum circuit denoted $\mathcal{C}$. Since we are assuming that the verifier has a constant-size quantum device, the state $|\psi\rangle$ will be a product state, i.e., $|\psi\rangle = |\psi_1\rangle \otimes |\psi_2\rangle \otimes \ldots \otimes |\psi_n\rangle$. For simplicity, assume each $|\psi_i\rangle$ is one qubit, though any constant number of qubits is allowed. In Clifford-QAS VQC the verifier will use the prover as an untrusted quantum storage device. Specifically, each $|\psi_i\rangle$, from $|\psi\rangle$, will be paired with a constant-size flag system in the accept state, $|acc\rangle$, resulting in a *block* of the form $|block_i\rangle = |\psi_i\rangle |acc\rangle$. Each block will be encoded, by having a random Clifford operation applied on top of it. The verifier prepares these blocks, one at a time, for all $i \in \{1, \ldots, n\}$, and sends them to the prover. The prover is then asked to return pairs of blocks to the verifier so that she may apply gates from $\mathcal{C}$ on them (after undoing the Clifford operations). The verifier then applies new random Cliffords on the blocks and sends them back to the prover. The process continues until all gates in $\mathcal{C}$ have been applied.

But what if the prover corrupts the state or deviates in some way? This is where the QAS enters the picture. Since each block has a random Clifford operation applied, the idea is to have the verifier use the Clifford QAS scheme to ensure that the quantum state remains authenticated after each gate in the quantum circuit is applied. In other words, if the prover attempts to deviate at any point resulting in a corrupted state, this should be detected by the authentication scheme. Putting everything together, the protocol works as follows:

(1) Suppose the input state that the verifier intends to prepare is $|\psi\rangle = |\psi_1\rangle \otimes |\psi_2\rangle \otimes \ldots \otimes |\psi_n\rangle$, where each $|\psi_i\rangle$ is a one qubit state[18]. Also let $\mathcal{C}$ be quantum circuit that the verifier wishes to apply on $|\psi\rangle$. The verifier prepares (one block at a time) the state $|\psi\rangle |flag\rangle = |block_1\rangle \otimes |block_2\rangle \otimes \ldots \otimes |block_n\rangle$, where $|block_i\rangle = |\psi_i\rangle |acc\rangle$ and each $|acc\rangle$ state consists of a constant number m of qubits. Additionally let the size of each block be $t = m + 1$.

(2) The verifier applies a random Clifford operation, from the set $\mathfrak{C}_t$ on each block and sends it to the prover.

(3) The verifier requests a pair of blocks, $(|block_i\rangle, |block_j\rangle)$, from the prover, in order to apply a gate from $\mathcal{C}$ on the corresponding qubits, $(|\psi_i\rangle, |\psi_j\rangle)$. Once the blocks have been received, the verifier undoes the random Cliffords and measures the flag registers, aborting if these are not in the $|acc\rangle$ state. Otherwise, the verifier performs the gate from $\mathcal{C}$, applies new random Cliffords on each block and sends them back to the prover. This step repeats until all gates in $\mathcal{C}$ have been performed.

(4) Once all gates have been performed, the verifier requests all the blocks (one by one) in order to measure the output. As in the previous step, the verifier

[18]This can simply be the state $|x\rangle$, if the verifier wishes to apply $\mathcal{C}$ on the classical input x. However, the state can be more general which is why we are not restricting it to be $|x\rangle$.

will undo the Clifford operations first and measure the flag registers, aborting if any of them are not in the $|acc\rangle$ state.

We can see that the security of this protocol reduces to the security of the Clifford QAS. Moreover, it is also clear that if the prover behaves honestly, then the verifier will obtain the correct output state exactly. Hence:

Theorem 1. *For a fixed constant $m > 0$, Clifford-QAS VQC is a prepare-and-send* QPIP *protocol having correctness $\delta = 1$ and verifiability $\epsilon = 2^{-m}$.*

2.1.2. Poly-QAS VQC. The second protocol in [26, 27], is referred to as *Polynomial QAS-based Verifiable Quantum Computing* (Poly-QAS VQC). It improves upon the previous protocol by removing the interactive quantum communication between the verifier and the prover, reducing it to a single round of quantum messages sent at the beginning of the protocol. To encode the input, this protocol uses a specific type of quantum error correcting code known as a *polynomial CSS code* [54]. We will not elaborate on the technical details of these codes as that is beyond the scope of this review. We only mention a few basic characteristics which are necessary in order to understand the Poly-QAS VQC protocol. The polynomial CSS codes operate on *qudits* instead of qubits. A q-qudit is simply a quantum state in a q-dimensional Hilbert space. The generalized computational basis for this space is given by $\{|i\rangle\}_{i\leq q}$. The code takes a q-qudit, $|i\rangle$, as well as $|0\rangle$ states, and encodes them into a state of $t = 2d + 1$ qudits as follows:

$$E\,|i\rangle\,|0\rangle^{\otimes t-1} = \sum_{p, deg(p)\leq d, p(i)=0} |p(\alpha_1)\rangle\,|p(\alpha_2)\rangle \ldots |p(\alpha_t)\rangle\,, \tag{27}$$

where E is the encoding unitary, p ranges over polynomials of degree less than d over the field F_q of integers *mod* q, and $\{\alpha_j\}_{j\leq t}$ is a fixed set of m *non-zero* values from F_q (it is assumed that $q > t$). The code can detect errors on at most d qudits and can correct errors on up to $\lfloor \frac{d}{2} \rfloor$ qudits (hence $\lfloor \frac{d}{2} \rfloor$ is the weight of the code). Importantly, the code is transversal for Clifford operations. Aharonov et al consider a slight variation of this scheme called a *signed polynomial code*, which allows one to randomize over different polynomial codes. The idea is to have the encoding (and decoding) procedure also depend on a key $k \in \{-1, +1\}^t$ as follows:

$$E_k\,|i\rangle\,|0\rangle^{\otimes t-1} = \sum_{p, deg(p)\leq d, p(i)=0} |k_1 p(\alpha_1)\rangle\,|k_2 p(\alpha_2)\rangle \ldots |k_t p(\alpha_t)\rangle\,. \tag{28}$$

The signed polynomial CSS code can be used to create a simple authentication scheme having security $\epsilon = 2^{-d}$. This works by having the sender encode the state $|\Psi_{in}\rangle = |\psi\rangle\,|0\rangle^{\otimes t-1}$, where $|\psi\rangle$ is a qudit to be authenticated, in the signed code and then one-time padding the encoded state. Note that the $|0\rangle^{\otimes t-1}$ part of the state is acting as a flag system. We are assuming that the sender and the receiver share both the sign key of the code and the key for the one-time padding. The one-time padded state is then sent over the insecure channel. The receiver undoes the pad and applies the inverse of the encoding operation. It then measures the last $t-1$ qudits, accepting if and only if they are all in the $|0\rangle$ state. Proving security is similar to the Clifford QAS and relies on two results:

Lemma 2 (Pauli twirl). *Let P_1, P_2 be two operators from the n-qudit Pauli group, denoted $\mathbb{P}_n$, such that $P_1 \neq P_2$. For any n-qudit density matrix ρ it is the case that:*

$$\sum_{Q \in \mathbb{P}_n} Q^\dagger P_1 Q \rho Q^\dagger P_2 Q = 0 \,. \tag{29}$$

This result is identical to the Clifford twirl lemma, except the Clifford operations are replaced with Paulis[19]. The result is also valid for qubits.

Lemma 3 (Signed polynomial code security). *Let $\rho = |\psi\rangle\langle\psi| \otimes |0\rangle\langle 0|^{\otimes t-1}$, be a state which will be encoded in the signed polynomial code, $P = (I - |\psi\rangle\langle\psi|) \otimes |0\rangle\langle 0|^{\otimes t-1}$, be a projector onto the orthogonal complement of $|\psi\rangle$ and on $|0\rangle^{t-1}$, and $Q \in \mathbb{P}_t \backslash \{I\}$ be a non-identity Pauli operation on t qudits. Then it is the case that:*

$$\frac{1}{2^t} \sum_{k \in \{-1,+1\}^t} Tr\left(P \; E_k^\dagger Q E_k \; \rho \; E_k^\dagger Q E_k\right) \leq \frac{1}{2^{t-1}} \,. \tag{30}$$

Using these two results, and the ideas from the Clifford QAS scheme, it is not difficult to prove the security of the above described authentication scheme. As before, the eavesdropper's map is decomposed into Kraus operators which are then expanded into Pauli operations. Since the sender's state is one-time padded, the Pauli twirl lemma will turn the eavesdropper's deviation into a convex combination of Pauli deviations:

$$\frac{1}{2^t} \sum_{k \in \{-1,+1\}^t} \sum_{Q \in \mathbb{P}_t} \beta_Q \; Q E_k |\Psi_{in}\rangle\langle\Psi_{in}| E_k^\dagger Q^\dagger, \tag{31}$$

which can be split into the identity and non-identity Paulis:

$$\frac{1}{2^t} \sum_{k \in \{-1,+1\}^t} \left(\beta_I E_k |\Psi_{in}\rangle\langle\Psi_{in}| E_k^\dagger + \sum_{Q \in \mathbb{P}_t \backslash \{I\}} \beta_Q \; Q E_k |\Psi_{in}\rangle\langle\Psi_{in}| E_k^\dagger Q^\dagger \right), \tag{32}$$

where β_Q are positive real coefficients satisfying:

$$\sum_{Q \in \mathbb{P}_n} \beta_Q = 1 \,. \tag{33}$$

The receiver takes this state and applies the inverse encoding operation, resulting in:

$$\rho = \frac{1}{2^t} \sum_{k \in \{-1,+1\}^t} \left(\beta_I |\Psi_{in}\rangle\langle\Psi_{in}| + \sum_{Q \in \mathbb{P}_t \backslash \{I\}} \beta_Q \; Q E_k |\Psi_{in}\rangle\langle\Psi_{in}| E_k^\dagger Q^\dagger \right) \,. \tag{34}$$

[19]Note that by abuse of notation we assume $\mathbb{P}_n$ refers to the group of generalized Pauli operations over qudits, whereas, typically, one uses this notation to refer to the Pauli group of qubits.

But now we know that $\epsilon = Tr(P_{incorrect}\rho)$, and using Lemma 3 together with the facts that $Tr(P_{incorrect} |\Psi_{in}\rangle \langle\Psi_{in}|) = 0$ and that the β_Q coefficients sum to 1 we end up with:

$$\epsilon \leq \frac{1}{2^{t-1}} \leq \frac{1}{2^d} \,. \tag{35}$$

There are two more aspects to be mentioned before giving the steps of the Poly-QAS VQC protocol. The first is that the encoding procedure for the signed polynomial code is implemented using the following interpolation operation:

$$D_k |i\rangle |k_2 p(\alpha_2)\rangle \ldots |k_{d+1} p(\alpha_{d+1})\rangle |0\rangle^{\otimes d} = |k_1 p(\alpha_1)\rangle \ldots |k_t p(\alpha_t)\rangle \,. \tag{36}$$

The inverse operation $D_k^\dagger$ can be though of as a decoding of one term from the superposition in Equation 28. Akin to Lemma 3, the signed polynomial code has the property that, when averaging over all sign keys, k, if such a term had a non-identity Pauli applied to it, when decoding it with $D_k^\dagger$, the probability that its last d qudits are not $|0\rangle$ states is upper bounded by 2^{-d}.

The second aspect is that, as mentioned, the signed polynomial code is transversal for Clifford operations. However, in order to apply non-Clifford operations it is necessary to measure encoded states together with so-called *magic states* (which will also be encoded). This manner of performing gates is known as *gate teleportation* [55]. The target state, on which we want to apply a non-Clifford operation, and the magic state are first entangled using a Clifford operation and then the magic state is measured in the computational basis. The effect of the measurement is to have a non-Clifford operation applied on the target state, along with Pauli errors which depend on the measurement outcome. For the non-Clifford operations, Aharonov et al use Toffoli gates[20].

Given all of these, the Poly-QAS VQC protocol works as follows:

(1) Suppose the input state that the verifier intends to prepare is $|\psi\rangle = |\psi_1\rangle \otimes |\psi_2\rangle \otimes \ldots \otimes |\psi_n\rangle$, where each $|\psi_i\rangle$ is a q-qudit. Also suppose that the verifier wishes to apply the quantum circuit $\mathcal{C}$ on $|\psi\rangle$, which contains L Toffoli gates. The verifier prepares the state $|\Psi_{in}\rangle = |\psi_1\rangle |0\rangle^{t-1} \otimes |\psi_2\rangle |0\rangle^{t-1} \otimes \ldots \otimes |\psi_n\rangle |0\rangle^{t-1} \otimes |M_1\rangle |0\rangle^{3t-3} \otimes \ldots \otimes |M_L\rangle |0\rangle^{3t-3}$, where $t = 2d+1$ and each $|M_i\rangle$ is a 3-qudit magic state, used for performing Toffoli gates. Groups of t qubits will comprise a block as follows. The first n blocks are simply $|block_i\rangle = |\psi_i\rangle |0\rangle^{t-1}$, with $i \in \{1, \ldots, n\}$. Next, we have the states of the form $|M_i\rangle |0\rangle^{3t-3}$ which consist of 3 blocks, each. Each block, from such a state, will comprise of one qudit from $|M_i\rangle$ and a $|0\rangle^{t-1}$ state. Note that we can no longer represent these blocks as pure states, since the 3 qudits of a $|M_i\rangle$ state are entangled. So, to summarize, each block contains one qudit from either the state $|\psi\rangle$ or a magic state $|M_i\rangle$, together with a flag system, $|0\rangle^{t-1}$.

[20]See Subsection 7.1.6 for the definition of the Toffoli gate.

(2) The verifier encodes each block in a signed polynomial code with a randomly chosen key $k \in \{-1, +1\}^t$ (the same key for each block) and then quantum one-time pads each block (using different keys for the padding of each block). The blocks are prepared and encoded in sequence (the verifier has the ability to process 3 blocks, or $3t$ qudits, at a time) and then sent to the prover.

(3) When applying Clifford operations, the verifier simply asks the prover to apply the gates in a transversal fashion. Since Cliffords commute with Paulis, the verifier then updates the one-time pad keys similar to Childs' protocol (see Subsection 1.1.2).

(4) When applying a Toffoli gate, the verifier asks the prover to measure 3 blocks, comprising a magic state, in the computational basis and report the measurement outcomes. It is assumed that the magic state was entangled, using a Clifford operation, with 3 target blocks on which the Toffoli gate is to be applied. The verifier undoes the (classical) one-time padding of the measurement outcomes and expects each of the 3 groups of measurement outcomes (associated with each of the 3 blocks) to be of the form $[k_1 p(\alpha_1), \ldots, k_t p(\alpha_t)]$. The verifier then takes these classical strings and turns them into states of the form $|\phi\rangle = |k_1 p(\alpha_1)\rangle \ldots |k_t p(\alpha_t)\rangle$ (using her constant-sized quantum computer)[21]. She then applies $D_k^\dagger$ on each of these $|\phi\rangle$ states and checks that the last d qudits, of each state, are $|0\rangle$, aborting otherwise. Assuming not-abort, the verifier instructs the prover to perform the appropriate Pauli corrections resulting from the gate teleportation.

(5) Once all gates have been performed, the verifier instructs the prover to measure all blocks in the computational basis. As in step 4, the verifier will then de-one-time pad the outcomes, apply $D_k^\dagger$ to each state of the form $|\phi\rangle$ (prepared from these outcomes), and check that the last d qudits are $|0\rangle$, aborting otherwise.

The protocol is schematically illustrated in Figure 5.

As with the previous protocol, the security is based on the security of the authentication scheme. However, there is a significant difference. In the Clifford-QAS VQC protocol, one could always assume that the state received by the verifier was the correctly encoded state with a deviation on top that was independent of this encoding. However, in the Poly-QAS VQC protocol, the quantum state is never returned to the verifier and, moreover, the prover's instructed actions on this state are adaptive based on the responses of the verifier. Since the prover is free to deviate at any point throughout the protocol, if we try to commute all of his deviations to the end (i.e., view the output state as the correct state resulting from an honest run of the protocol, with a deviation on top that is independent of the secret parameters), we find that the output state will have a deviation on top

[21]Note that no actual quantum state was returned to the verifier by the prover. Instead, she locally prepared a quantum state from the classical outcomes reported by the prover.

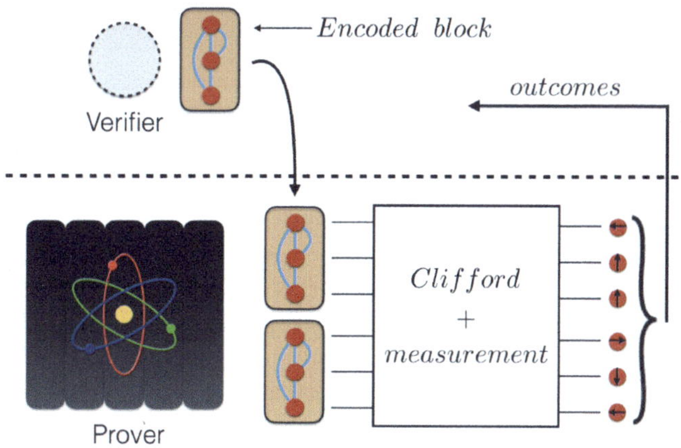

FIGURE 5. Poly-QAS VQC

which depends on the verifier's responses. Since the verifier's responses depend on the secret keys, we cannot directly use the security of the authentication scheme to prove that the protocol is 2^{-d}-verifiable.

The solution, as explained in [27], is to consider the state of the entire protocol comprising of the prover's system, the verifier's system *and* the transcript of all classical messages exchanged during the protocol. For a fixed interaction transcript, the prover's attacks can be commuted to the end of the protocol. This is because, if the transcript is fixed, there is no dependency of the prover's operations on the verifier's messages. We simply view all of his operations as unitaries acting on the joint system of his private memory, the input quantum state and the transcript. One can then use Lemma 2 and Lemma 3 to bound the projection of this state onto the incorrect subspace with acceptance. The whole state, however, will be a mixture of all possible interaction transcripts, but since each term is bounded and the probabilities of the terms in the mixture must add up to one, it follows that the protocol is 2^{-d}-verifiable:

Theorem 2. *For a fixed constant $d > 0$, Poly-QAS VQC is a prepare-and-send* QPIP *protocol having correctness $\delta = 1$ and verifiability $\epsilon = 2^{-d}$.*

Before ending this subsection, let us briefly summarize the two protocols in terms of the verifier's resources. In both protocols, if one fixes the security parameter, ϵ, the verifier must have a $O(\log(1/\epsilon))$-size quantum computer. Additionally, both protocols are interactive with the total amount of communication (number of messages times the size of each message) being upper bounded by $O(|\mathcal{C}| \cdot \log(1/\epsilon))$, where $\mathcal{C}$ is the quantum circuit to be performed[22]. However, in Clifford-QAS VQC, this communication is quantum whereas in Poly-QAS VQC

[22]To be precise, the communication in the Poly-QAS VQC scheme is $O((n+L) \cdot \log(1/\epsilon))$, where n is the size of the input and L is the number of Toffoli gates in $\mathcal{C}$.

only one quantum message is sent at the beginning of the protocol and the rest of the interaction is classical.

2.2. Trap-Based Verification

In this subsection we discuss *Verifiable Universal Blind Quantum Computing* (VUBQC), which was developed by Fitzsimons and Kashefi in [28]. The protocol is written in the language of MBQC and relies on two essential ideas. The first is that an MBQC computation can be performed blindly, using UBQC, as described in Subsection 1.1. The second is the idea of embedding checks or *traps* in a computation in order to verify that it was performed correctly. Blindness will ensure that these checks remain hidden and so any deviation by the prover will have a high chance of triggering a trap. Notice that this is similar to the QAS-based approaches where the input state has a flag subsystem appended to it in order to detect deviations and the whole state has been encoded in some way so as to hide the input and the flag subsystem. This will lead to a similar proof of security. However, as we will see, the differences arising from using MBQC and UBQC lead to a reduction in the quantum resources of the verifier. In particular, in VUBQC the verifier requires only the ability to prepare single qubit states, which will be sent to the prover, in contrast to the QAS-based protocols which required the verifier to have a constant-size quantum computer.

Recall the main steps for performing UBQC. The client, Alice, sends qubits of the form $|+_{\theta_i}\rangle$ to Bob, the server, and instructs him to entangle them according to a graph structure, G, corresponding to some universal graph state. She then asks him to measure qubits in this graph state at angles $\delta_i = \phi'_i + \theta_i + r_i\pi$, where ϕ'_i is the corrected computation angle and $r_i\pi$ acts a random Z operation which flips the measurement outcome. Alice will use the measurement outcomes, denoted b_i, provided by Bob to update the computation angles for future measurements. Throughout the protocol, Bob's perspective is that the states, measurements and measurement outcomes are indistinguishable from random. Once all measurements have been performed, Alice will undo the r_i padding of the final outcomes and recover her output. Of course, UBQC does not provide any guarantee that the output she gets is the correct one, since Bob could have deviated from her instructions.

Transitioning to VUBQC, we will identify Alice as the verifier and Bob as the prover. To augment UBQC with the ability to detect malicious behaviour on the prover's part, the verifier will introduce traps in the computation. How will she do this? Recall that the qubits which will comprise $|G\rangle$ need to be entangled with the CZ operation. Of course, for XY-plane states CZ does indeed entangle the states. However, if either qubit, on which CZ acts, is $|0\rangle$ or $|1\rangle$, then no entanglement is created. So suppose that we have a $|+_\theta\rangle$ qubit whose neighbours, according to G, are computational basis states. Then, this qubit will remain disentangled from the rest of the qubits in $|G\rangle$. This means that if the qubit is measured at its preparation angle, the outcome will be deterministic. The verifier can exploit this fact to certify that the prover is performing the correct measurements. Such states are referred to as *trap qubits*, whereas the $|0\rangle$, $|1\rangle$ neighbours are referred to as *dummy qubits*.

Importantly, as long as G's structure remains that of a universal graph state[23] and as long as the dummy qubits and the traps are chosen at random, adding these extra states as part of the UBQC computation will not affect the blindness of the protocol. The implication of this is that the prover will be completely unaware of the positions of the traps and dummies. The traps effectively play a role that is similar to that of the flag subsystem in the authentication-based protocols. The dummies, on the other hand, are there to ensure that the traps do not get entangled with the rest of qubits in the graph state. They also serve another purpose. When a dummy is in a $|1\rangle$ state, and a CZ acts on it and a trap qubit, in the state $|+_\theta\rangle$, the effect is to "flip" the trap to $|-_\theta\rangle$ (alternatively $|-_\theta\rangle$ would have been flipped to $|+_\theta\rangle$). This means that if the trap is measured at its preparation angle, θ, the measurement outcome will also be flipped, with respect to the initial preparation. Conversely, if the dummy was initially in the state $|0\rangle$, then no flip occurs. Traps and dummies, therefore, serve to also certify that the prover is performing the CZ operations correctly. Thus, by using the traps (and the dummies), the verifier can check both the prover's measurements and his entangling operations and hence verify his MBQC computation.

We are now ready to present the steps of VUBQC:

(1) The verifier chooses an input x and a quantum computation $\mathcal{C}$ that she would like the prover to perform on $|x\rangle$[24].

(2) She converts x and $\mathcal{C}$ into a pair $(G, \{\phi_i\}_i)$, where $|G\rangle$ is an N-qubit universal graph state (with an established ordering for measuring the qubits), which admits an embedding of T traps and D dummies. We therefore have that $N = T + D + Q$, where $Q = O(|\mathcal{C}|)$ is the number of computation qubits used for performing $\mathcal{C}$ and $\{\phi_i\}_{i \leq Q}$ is the associated set of computation angles[25].

(3) Alice picks, uniformly at random, values θ_i, with i going from 1 to $T + Q$, from the set $\{0, \pi/4, 2\pi/4, \ldots, 7\pi/4\}$ as well as values r_i from the set $\{0, 1\}$ for the trap and computation qubits.

(4) She then prepares the $T + Q$ states $|+_{\theta_i}\rangle$, as well as D dummy qubits which are states chosen at random from $\{|0\rangle, |1\rangle\}$. All these states are sent to Bob, who is instructed to entangle them, using CZ operations, according to the graph structure G.

---

[23]Note that adding dummy qubits into the graph will have the effect of disconnecting qubits that would otherwise have been connected. It is therefore important that the chosen graph state allows for the embedding of traps and dummies so that the desired computation can still be performed. For instance, the brickwork state from Subsection 7.2 allows for only one trap qubit to be embedded, whereas other graph states allows for multiple traps. See [28,56] for more details.

[24]As in the previous protocols, this need not be a classical input and the verifier could prepare an input of the form $|\psi\rangle = |\psi_1\rangle \otimes \ldots \otimes |\psi_n\rangle$.

[25]Note that the number of traps, T, and the number of dummies, D, are related, since each trap should have only dummy neighbours in $|G\rangle$.

(5) Alice then asks Bob to measure the qubits as follows: computation qubits will be measured at $\delta_i = \phi'_i + \theta_i + r_i\pi$, where ϕ'_i is an updated version of ϕ_i that incorporates corrections resulting from previous measurements; trap qubits will be measured at $\delta_i = \theta_i + r_i\pi$; dummy qubits are measured at randomly chosen angles from $\{0, \pi/4, 2\pi/4, \ldots, 7\pi/4\}$. This step is interactive as Alice needs to update the angles of future measurements based on past outcomes. The number of rounds of interaction is proportional to the depth of $\mathcal{C}$. If any of the trap measurements produce incorrect outcomes, Alice will abort upon completion of the protocol.

(6) Assuming all trap measurements succeeded, after all the measurements have been performed, Alice undoes the r_i one-time padding of the measurement outcomes, thus recovering the outcome of the computation.

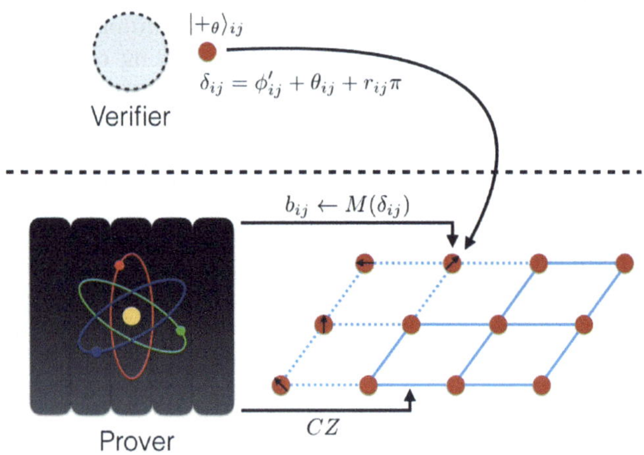

FIGURE 6. Verifiable Universal Blind Quantum Computing

The protocol is illustrated schematically in Figure 6, where all the parameters have been labelled by their position, (i, j), in a rectangular cluster state.

One can see that VUBQC has correctness $\delta = 1$, since if the prover behaves honestly then all trap measurements will produce the correct result and the computation will have been performed correctly. What about verifiability? We will first answer this question for the case where there is a single trap qubit ($T = 1$) at a uniformly random position in $|G\rangle$, denoted $|+_{\theta_t}\rangle$. Adopting a similar notation to that from [28], we let:

$$\mathcal{B}_j(\nu) = \sum_{\mathbf{s}} p_{\nu,j}(\mathbf{s}) \, |\mathbf{s}\rangle \langle \mathbf{s}| \otimes \rho^{\mathbf{s}}_{\nu,j} \tag{37}$$

denote the outcome density operator of *all classical and quantum messages* exchanged between the verifier and the prover throughout the protocol, excluding the last round of measurements (which corresponds to measuring the output of the computation). Additionally, ν denotes the set of secret parameters of Alice (i.e., the positions of the traps and dummies as well as the sets $\{\phi_i\}_i$, $\{\theta_i\}_i$ and $\{r_i\}_i$); j ranges over the possible strategies of the prover[26] with $j = 0$ corresponding to the honest strategy; $\mathbf{s}$ is a binary vector which ranges over all possible *corrected* values of the measurement outcomes sent by the prover; lastly, $\rho^{\mathbf{s}}_{\nu,j}$ is the state of the unmeasured qubits, representing the output state of the computation (prior to the final measurement). To match Definition 2, one also considers:

$$P^{\nu}_{incorrect} = (I - \mathcal{C} \left|x\right\rangle \left\langle x\right| \mathcal{C}^{\dagger}) \otimes \left|+^{\nu}_{\theta_t}\right\rangle \left\langle +^{\nu}_{\theta_t}\right| \tag{38}$$

be the projection onto the orthogonal complement of the correct output together with the trap state being projected onto acceptance. The dependence on ν, for the trap qubit, arises because the acceptance outcome depends on the states of the dummy neighbors for that qubit. This is because if one of the dummies is $\left|1\right\rangle$, the CZ operation has the effect of flipping $\left|+_{\theta_t}\right\rangle$ to $\left|-_{\theta_t}\right\rangle$. Additionally, ν also encodes the position of this trap, in the graph state, as well as the Z flip specified by the r_i parameter, for $i = t$. One then needs to find an ϵ such that:

$$Tr\left(\sum_{\nu} p(\nu) P^{\nu}_{incorrect} \mathcal{B}_j(\nu)\right) \leq \epsilon\,. \tag{39}$$

This is done in a manner similar to the proof of security for the Poly-QAS VQC scheme of the previous section[27]. Specifically, one fixes the interaction transcript for the protocol. In this case, any deviation that the prover performs is independent of the secret parameters of the verifier and can therefore be commuted to the end of the protocol. The outcome density operator $\mathcal{B}_j(\nu)$ can then be expressed as the ideal outcome with a deviation, $\mathcal{E}_j$, on top that is independent of ν:

$$\mathcal{B}_j(\nu) = \mathcal{E}_j(\mathcal{B}_0(\nu))\,. \tag{40}$$

The deviation $\mathcal{E}_j$ is then decomposed into Kraus operators which, in turn, are decomposed into Pauli operators leading to:

$$\mathcal{B}_j(\nu) = \sum_{P_i, P_k \in \mathbb{P}_N} \alpha_{ik}(j)\; P_i \mathcal{B}_0(\nu) P_k\,, \tag{41}$$

where $\alpha_{ik}(j)$ are the complex coefficients for the Pauli operators. The UBQC protocol ensures that, when summing over the secret parameters ν, the qubits sent by the verifier to the prover are effectively one-time padded. This means that one can use a version of the Pauli twirl Lemma 2 and get that:

$$\sum_{\nu} p(\nu) \mathcal{B}_j(\nu) = \sum_{\nu} p(\nu) \sum_{P_i \in \mathbb{P}_N} |\alpha_{ii}(j)|^2\; P_i \mathcal{B}_0(\nu) P_i\,. \tag{42}$$

---

[26]Since the prover is unbounded and is free to choose any of the *uncountably* many CPTP strategies, j should be thought more of as a symbolic parameter indicating that there is a dependence on the prover's strategy and whether or not this strategy is the ideal one.

[27]Note that the security proof for Poly-QAS VQC was in fact inspired from that of the VUBQC protocol, as mentioned in [27].

In other words, the resulting state, averaged over the secret parameters, is a convex combination of Pauli deviations. We are only interested in non-trivial deviations, i.e., deviations that can corrupt the output (and hence also flip the trap). The position of the trap qubit is randomized, so that it is equally likely that any of the N qubits is the trap. But if this is the case and we assume that non-trivial deviations have a non-zero weight in the convex combination, then one gets that the probability for the trap to be flipped is lower bounded by the *uniform* probability of $1/N$. Hence, for the case of a single trap qubit one has $\epsilon = 1 - \frac{1}{N}$.

If however, there are multiple trap states, the bound improves. Specifically, for a type of resource state called *dotted-triple graph*, the number of traps can be a constant fraction of the total number of qubits, yielding $\epsilon = 8/9$. If the protocol is then repeated a constant number of times, d, with the verifier aborting if any of these runs gives incorrect trap outcomes, it can be shown that $\epsilon = (8/9)^d$ [56]. Alternatively, if the input state and computation are encoded in an error correcting code of distance d, then one again obtains $\epsilon = (8/9)^d$. This is useful if one is interested in a quantum output, or a classical bit string output. If, instead, one would only like a single bit output (i.e., the outcome of the decision problem) then sequential repetition and taking the majority outcome is sufficient. The fault tolerant encoding need not be done by the verifier. Instead, the prover will simply be instructed to prepare a larger resource state which also offers topological error-correction. See [28, 57, 58] for more details. An important observation, however, is that the fault tolerant encoding, just like in the Poly-QAS VQC protocol, is used *only to boost security* and not for correcting deviations arising from faulty devices. This latter case is discussed in Section 5.2. To sum up:

Theorem 3. *For a fixed constant $d > 0$, VUBQC is a prepare-and-send* QPIP *protocol having correctness $\delta = 1$ and verifiability $\epsilon = (8/9)^d$.*

It should be noted that in the original construction of the protocol, the fault tolerant encoding, used for boosting security, required the use of a resource state having $O(|\mathcal{C}|^2)$ qubits. The importance of the dotted-triple graph construction is that it achieves the same level of security while keeping the number of qubits linear in $|\mathcal{C}|$. The same effect is achieved by a composite protocol which combines the Poly-QAS VQC scheme, from the previous section, with VUBQC [52]. This works by having the verifier run small instances of VUBQC in order to prepare the encoded blocks used in the Poly-QAS VQC protocol. Because of the blindness property, the prover does not learn the secret keys used in the encoded blocks. The verifier can then run the Poly-QAS VQC protocol with the prover, using those blocks. This hybrid approach illustrates how composition can lead to more efficient protocols. In this case, the composite protocol maintains a single qubit preparation device for the verifier (as opposed to a $O(\log(1/\epsilon))$-size quantum computer) while also achieving linear communication complexity. We will encounter other composite protocols when reviewing entanglement-based protocols in Section 4.

Lastly, let us explicitly state the resources and overhead of the verifier throughout the VUBQC protocol. As mentioned, the verifier requires only a single-qubit preparation device, capable of preparing states of the form $|+_\theta\rangle$, with $\theta \in \{0, \pi/4, 2\pi/4, \ldots, 7\pi/4\}$, and $|0\rangle$, $|1\rangle$. The number of qubits needed is on the order of $O(|\mathcal{C}|)$. After the qubits have been sent to the prover, the two interact classically and the size of the communication is also on the order of $O(|\mathcal{C}|)$.

2.3. Verification Based on Repeated Runs

The final prepare-and-send protocol we describe is the one defined by Broadbent in [29]. While the previous approaches relied on hiding a flag subsystem or traps in either the input or the computation, this protocol has the verifier alternate between different runs designed to either test the behaviour of the prover or perform the desired quantum computation. We will refer to this as the *Test-or-Compute* protocol. From the prover's perspective, the possible runs are indistinguishable from each other, thus making him unaware if he is being tested or performing the verifier's chosen computation. Specifically, suppose the verifier would like to delegate the quantum circuit $\mathcal{C}$ to be applied on the $|0\rangle^{\otimes n}$ state[28], where n is the size of the input. The verifier then chooses randomly between three possible runs:

- **Computation run**. The verifier delegates $\mathcal{C}\,|0\rangle^{\otimes n}$ to the prover.
- **X-test run**. The verifier delegates the identity computation on the $|0\rangle^{\otimes n}$ state to the prover.
- **Z-test run**. The verifier delegates the identity computation on the $|+\rangle^{\otimes n}$ state to the prover.

It turns out that this suffices in order to test against any possible malicious behavior of the prover, with high probability.

In more detail, the protocol uses a technique for quantum computing on encrypted data, described in [59], which is similar to Childs' protocol from Subsection 1.1.2, except it does not involve two-way quantum communication. The verifier will one-time pad either the $|0\rangle^{\otimes n}$ state or the $|+\rangle^{\otimes n}$ state and send the qubits to the prover. The prover is then instructed to apply the circuit $\mathcal{C}$, which consists of the gates X, Z, H, T, CNOT. As we know, the Clifford operations commute with the one-time pad, so the verifier would only need to appropriately update the one-time pad to account for this. However, T gates do not commute with the pad. In particular, commuting them past the X gates introduces unwanted S operations. To resolve this issue, the verifier will use a particular gadget which will allow the prover to apply T and correct for S at the same time. This gadget is shown in Figure 7, reproduced from [29].

[28] The preparation of a specific input $|x\rangle$ can be done as part of the circuit $\mathcal{C}$.

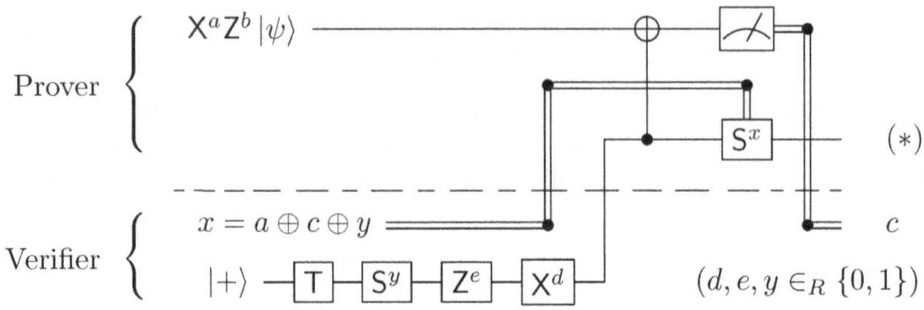

FIGURE 7. Gadget for performing T on one-time padded state, reproduced from [29]. Here $(*) = X^{a \oplus c} Z^{(a \oplus c) \cdot (d \oplus y) \oplus a \oplus b \oplus c \oplus e \oplus y} T |\psi\rangle$

The gadget works in a gate teleportation fashion. For each qubit, labelled j, on which the prover should apply a T gate, the verifier sends a qubit of the form $X^d Z^c S^y T |+\rangle$, as well as the classical bit $x = a \oplus c \oplus y$, where a is the X padding of qubit j and c, d and y are chosen at random. The verifier then instructs the prover to apply a CNOT between the sent qubit and qubit j, effectively entangling them, and then measure qubit j in the computational basis. Lastly, the verifier instructs the prover to apply an S^x gate to the sent qubit. The end result is that this qubit will be the same as the de-one-time padded qubit j but with a T and a new one-time pad acting on it. Importantly, the new pad is kept secret from the prover.

The T gate gadget allows the verifier to control the application of either a non-Clifford operation or a Clifford operation through gate teleportation. For note that if the verifier does not apply a T gate on the qubit sent to the prover, the resulting computation is Clifford. This is what allows the verifier to switch between the computation run and the two test runs. The prover cannot distinguish between the two cases, since his side of the gadget is identical in both instances. Thus, in a test run, the computation the prover performs will be Clifford and the verifier can simply update the one-time pad, of the input, accordingly. There is, however, one complication. In an X-test run, the input is $|0\rangle^{\otimes n}$ and should remain this way until the end of the circuit, up to qubit flips resulting from the one-time pad. But any Hadamard gate in the circuit will map $|0\rangle$ to $|+\rangle$. The same is true for Z-test runs, where $|+\rangle$ states can be mapped to $|0\rangle$. To resolve this issue, Broadbent uses the following identities:

$$HT^2HT^2HT^2H = H\,, \tag{43}$$

$$HHHH = I\,. \tag{44}$$

The idea is to have the prover implement each Hadamard operation in C by applying four H gates alternating with $S = T^2$ gates. Each T^2 operation is performed by using the T gate gadget twice. When the verifier chooses a computation run, she will apply the T gates in the gadget and therefore, via Equation 43, this leads to a Hadamard operation. Conversely, in a rest run, no T gates are applied, hence, from Equation 44, no Hadamard operation will act on the target qubit. Since the output is always measured, by the prover, in the computational basis, in an X-test run the verifier simply checks that the de-one-time padded output is $|0\rangle^{\otimes n}$.

There is, in fact, an additional testing step being performed during an X-test run. Consider the T gadget for such a run in Figure 8, reproduced from [29].

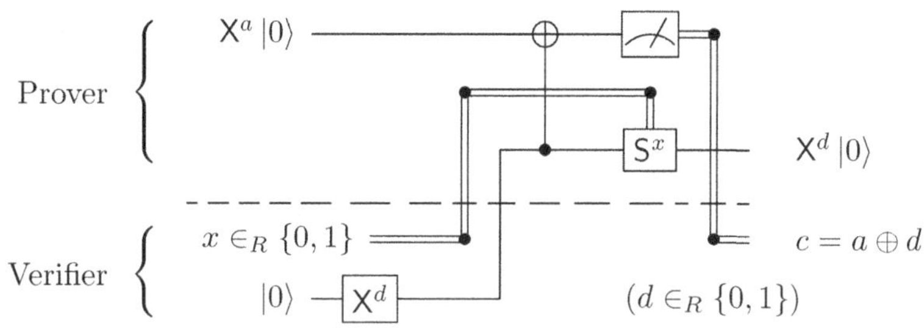

FIGURE 8. T gate gadget in X-test run, reproduced from [29]

Note that the measurement bit, c, provided by the prover to the verifier should be an *xor* of the original X padding of the input and the updated X padding of the input. Checking the value of this bit allows the verifier to test that the gadget was applied correctly.

What about the Z-test run? In that case, the output should be the $|+\rangle^{\otimes n}$ which, upon measurement, should collapse with equal probability into any of the 2^n possible binary strings. The verifier does not test for this. Instead, each H gate that should be applied by the prover has the effect of switching between the Z-test run and the X-test run. Thus, whenever a Hadamard is applied to a qubit, during a Z-test run, the verifier switches to an X-test run until the next Hadamard operation. In the X-test runs, the verifier will use the T gate gadget from Figure 8. These are the only checks that are performed in the Z-test run.

Putting everything together, the protocol starts by having the verifier choose an input x and a quantum computation $\mathcal{C}$ to act on this input (that includes preparing the input from the $|0\rangle^{\otimes n}$ state). Then, the verifier will randomly choose to perform one of the following:

- **Computation run**. The verifier one-time pads the $|0\rangle^{\otimes n}$ state and sends it to the prover. The prover is then instructed to apply $\mathcal{C}$ on this state, such that for each T gate in the circuit the prover and the verifier interact in order to perform the T gadget. Additionally, any H in $\mathcal{C}$ is performed as in Equation 43. For Clifford operations, the verifier updates the one-time pad of the state accordingly. The prover is instructed to measure the output state of the circuit in the computational basis and return the outcome to the verifier. The verifier undoes the padding of this outcome and accepts if the output of the circuit indicates acceptance.

- **X-test run**. The verifier one-time pads the $|0\rangle^{\otimes n}$ state and sends it to the prover. As in the computation run, for each T, the verifier and the prover will interact to run the T gate gadget. In this case, however, the verifier will use the T gate gadget from Figure 8, making the circuit effectively act as identity and checking that the prover is performing these gadgets correctly (rejecting otherwise). Additionally, the H gates in C will also act as identity, from Equation 44, as described previously. The verifier updates the one-time padding of the state accordingly for all gates in the circuit. Once the circuit is finished, the prover is instructed to measure the output in the computational basis and report the outcome to the verifier. The verifier accepts if the de-one-time padded output is $|0\rangle^{\otimes n}$.

- **Z-test run**. The verifier one-time pads the $|+\rangle^{\otimes n}$ state and sends it to the prover. As in the X-test run, the T gate gadgets will act as identity. The H operations that the prover performs will temporarily switch the Z-test run into an X-test run, in which the verifier uses the gadget from Figure 8 to check that prover implemented it correctly. Any subsequent H will switch back to a Z-test run. Additionally, the verifier updates the one-time padding of the state accordingly for all gates in the circuit. The prover is instructed to measure the output in the computational basis and report the outcome to the verifier, however in this case the verifier discards the output.

The asymmetry between the X-test run and the Z-test run stems from the fact that the output is always measured in the computational basis. This means that an incorrect output is one which has been bit-flipped. In turn, this implies that only X and Y operations on the output will act as deviations, since Z effectively acts as identity on computational basis states. If the circuit C does not contain any Hadamard gates and hence, the computation takes place entirely in the computational basis, then the X-test is sufficient for detecting such deviations. However, when Hadamard gates are present, this is no longer the case since deviations can occur in the conjugate basis, $(|+\rangle, |-\rangle)$, as well. This is why the Z-test is necessary. Its purpose is to check that the prover's operations are performed correctly when switching to the conjugate basis. For this reason, a Hadamard gate will switch a Z-test run into an X-test run which provides verification using the T gate gadget.

In terms of the correctness of the protocol, we can see that if the prover behaves honestly then the correct outcome is obtained in the computation run and the verifier will accept the test runs, hence $\delta = 1$[29]. For verifiability, the analysis is similar to the previous protocols. Suppose that $|\psi\rangle$ is either the $|0\rangle^{\oplus n}$ or the $|+\rangle^{\oplus n}$ state representing the input. Additionally, assuming there are t T gates in C (including the ones used for performing the Hadamards), let $|\phi\rangle$ be the state of the t qubits that the verifier sends for the T gate gadgets. Then, the

[29]However, note that if the verifier chooses a test run, in the case where the prover is honest, this will lead to acceptance irrespective of the outcome of the decision problem. This is in contrast to the previous protocols in which the testing is performed at the same time as the computation and, when the test succeeds, the verifier outputs the result of the computation.

one-time padded state that the prover receives is:

$$\frac{1}{4^{n+t}} \sum_{Q \in \mathbb{P}_{n+t}} Q \left|\psi\right\rangle \left\langle\psi\right| \otimes \left|\phi\right\rangle \left\langle\phi\right| Q^{\dagger} . \tag{45}$$

The prover is then instructed to follow the steps of the protocol in order to run the circuit $\mathcal{C}$. Note that all of the operations that he is instructed to perform are Clifford operations. This is because any non-Clifford operation from $\mathcal{C}$ is performed with the T gate gadgets (which require only Cliffords) and the states from $|\phi\rangle$, prepared by the verifier. In following the notation from [29], we denote the honest action of the protocol as C. As in the previous protocols, the prover's deviation can be commuted to the end of the protocol, so that it acts on top of the correct state. After expressing the deviation map in terms of Pauli operators one gets:

$$\frac{1}{4^{n+t}} \sum_{P_i, P_j, Q \in \mathbb{P}_{n+t}} \alpha_i \alpha_j^* \; P_i Q C \left|\psi\right\rangle \left\langle\psi\right| \otimes \left|\phi\right\rangle \left\langle\phi\right| C^{\dagger} Q^{\dagger} P_j^{\dagger} . \tag{46}$$

Note that we have also commuted C past the one-time pad so that it acts on the state $|\psi\rangle |\phi\rangle$, rather than on the one-time padded versions of these states. This is possible precisely because C is a Clifford operation and therefore commutes with Pauli operations. One can then assume that the verifier performs the decryption of the padding before the final measurement, yielding:

$$\frac{1}{4^{n+t}} \sum_{P_i, P_j, Q \in \mathbb{P}_{n+t}} \alpha_i \alpha_j^* \; Q^{\dagger} P_i Q C \left|\psi\right\rangle \left\langle\psi\right| \otimes \left|\phi\right\rangle \left\langle\phi\right| C^{\dagger} Q^{\dagger} P_j^{\dagger} Q . \tag{47}$$

We now use the Pauli twirl from Lemma 2 to get:

$$\frac{1}{4^{n+t}} \sum_{P_i \in \mathbb{P}_{n+t}} |\alpha_i|^2 \; P_i C \left|\psi\right\rangle \left\langle\psi\right| \otimes \left|\phi\right\rangle \left\langle\phi\right| C^{\dagger} P_i^{\dagger} , \tag{48}$$

which is a convex combination of Pauli attacks acting on the correct output state. If we now denote M to be the set of non-benign Pauli attacks (i.e., attacks which do not act as identity on the output of the computation), then one of the test runs will reject with probability:

$$\sum_{P_i \in M} |\alpha_i|^2 . \tag{49}$$

This is because non-benign Pauli X or Y operations are detected by the X-test run, whereas non-benign Pauli Z operations are detected by the Z-test run. Since either test occurs with probability $1/3$, it follows that, the probability of the verifier accepting an incorrect outcome is at most $2/3$, hence $\epsilon = 2/3$.

Note that when discussing the correctness and verifiability of the Test-or-Compute protocol, we have slightly abused the terminology, since this protocol does not rigorously match the established definitions for correctness and verifiability that we have used for the previous protocols. The reason for this is the fact that in the Test-or-Compute protocol there is no additional flag or trap subsystem to indicate failure. Rather, the verifier detects malicious behaviour by alternating between different runs. It is therefore more appropriate to view the Test-or-Compute protocol simply as a QPIP protocol having a constant gap between completeness and soundness:

Theorem 4. *Test-or-Compute is a prepare-and-send* QPIP *protocol having completeness* 8/9 *and soundness* 7/9.

In terms of the verifier's quantum resources, we notice that, as with the VUBQC protocol, the only requirement is the preparation of single qubit states. All of these states are sent in the first round of the protocol, the rest of the interaction being completely classical.

2.4. Summary of Prepare-and-Send Protocols

The protocols, while different, have the common feature that they all use blindness or have the potential to be blind protocols. Out of the five presented protocols, only the Poly-QAS VQC and the Test-or-Compute protocols are not explicitly blind since, in both cases, the computation is revealed to the server. However, it is relatively easy to make the protocols blind by encoding the circuit into the input (which is one-time padded). Hence, one can say that all protocols achieve blindness.

This feature is essential in the proof of security for these protocols. Blindness combined with either the Pauli twirl Lemma 2 or the Clifford twirl Lemma 1 have the effect of reducing any deviation of the prover to a convex combination of Pauli attacks. Each protocol then has a specific way of detecting such an attack. In the Clifford-QAS VQC protocol, the convex combination is turned into a uniform combination and the attack is detected by a flag subsystem associated with a quantum authentication scheme. A similar approach is employed in the Poly-QAS VQC protocol, using a quantum authentication scheme based on a special type of quantum error correcting code. The VUBQC protocol utilizes trap qubits and either sequential repetition or encoding in an error correcting code to detect Pauli attacks. Finally, the Test-or-Compute protocol uses a hidden identity computation acting on either the $|0\rangle^{\otimes n}$ or $|+\rangle^{\otimes n}$ states, in order to detect the malicious behavior of the prover.

Because of these differences, each protocol will have different "quantum requirements" for the verifier. For instance, in the authentication-based protocols, the verifier is assumed to be a quantum computer operating on a quantum memory of size $O(\log(1/\epsilon))$, where ϵ is the desired verifiability of the protocol. In VUBQC and Test-or-Compute, however, the verifier only requires a device capable of preparing single-qubit states. Additionally, out of all of these protocols, only Clifford-QAS VQC requires 2-way quantum communication, whereas the other three require the verifier to send only one quantum message at the beginning of the protocol, while the rest of the communication is classical. These facts, together with the communication complexities of the protocols are shown in Table 1.

As mentioned, if we want to make the Poly-QAS VQC and Test-or-Compute protocols blind, the verifier will hide her circuit by incorporating it into the input. The input would then consist of an encoding of C and an encoding of x. The

Protocol	Verifier resources	Communication	2-way quantum comm.
Clifford-QAS VQC	$O(\log(1/\epsilon))$	$O(N \cdot \log(1/\epsilon))$	Y
Poly-QAS VQC	$O(\log(1/\epsilon))$	$O((n+L) \cdot \log(1/\epsilon))$	N
VUBQC	$O(1)$	$O(N \cdot \log(1/\epsilon))$	N
Test-or-Compute	$O(1)$	$O((n+T) \cdot \log(1/\epsilon))$	N

TABLE 1. Comparison of prepare-and-send protocols. If denote as $\mathcal{C}$ the circuit that the verifier wishes to delegate to the prover, and as x the input to this circuit, then $n = |x|$, $N = |\mathcal{C}|$. Additionally, T denotes the number of T gates in $\mathcal{C}$, L denotes the number of Toffoli gates in $\mathcal{C}$ and ϵ denotes the verifiability of the protocols. The second column refers to the verifier's *quantum* resources. The third column quantifies the *total* communication complexity, both classical and quantum, of the protocols (i.e., number of messages times the size of a message).

prover would be asked to perform controlled operations from the part of the input containing the description of $\mathcal{C}$, to the part containing x, effectively acting with $\mathcal{C}$ on x. We stress that in this case, the protocols would have a communication complexity of $O(|\mathcal{C}| \cdot \log(1/\epsilon))$, just like VUBQC and Clifford-QAS VQC[30].

3. Receive-and-Measure Protocols

The protocols presented so far have utilized a verifier with a trusted preparation device (and potentially a trusted quantum memory) interacting with a prover having the capability of storing and performing operations on arbitrarily large quantum systems. In this section, we explore protocols in which the verifier possesses a trusted measurement device. The point of these protocols is to have the prover prepare a specific quantum state and send it to the verifier. The verifier's measurements have the effect of either performing the quantum computation or extracting the outcome of the computation. An illustration of receive-and-measure protocols is shown in Figure 9.

For prepare-and-send protocols we saw that blindness was an essential feature for achieving verifiability. While most of the receive-and-measure protocols are blind as well, we will see that it is possible to perform verification without hiding any information about the input or computation, from the prover. Additionally, while in prepare-and-send protocols the verifier was sending an encoded or encrypted quantum state to the prover, in receive-and-measure protocols, the

[30]Technically, the complexity should be $O((|x| + |\mathcal{C}|) \cdot \log(1/\epsilon))$, however we are assuming that $\mathcal{C}$ acts non-trivially on x (i.e., there are at least $|x|$ gates in $\mathcal{C}$).

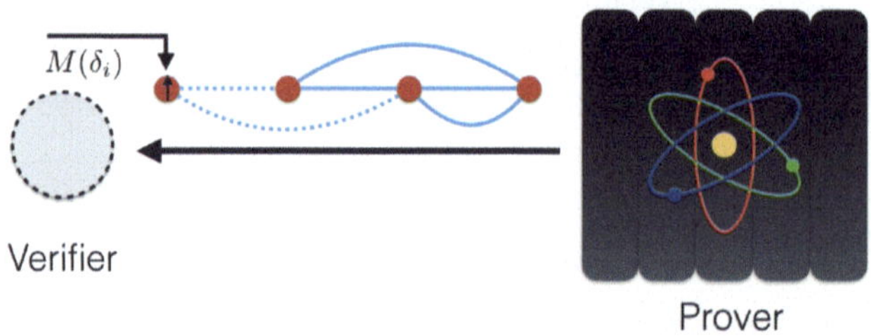

FIGURE 9. Receive-and-measure protocols

quantum state received by the verifier is not necessarily encoded or encrypted. Moreover, this state need not contain a flag or a trap subsystem. For this reason, we can no longer consistently define ϵ-verifiability and δ-correctness, as we did for prepare-and-send protocols. Instead, we will simply view receive-and-measure protocols as QPIP protocols.

The protocols presented in this section are:

1. **Subsection 3.1**: a *measurement-only* protocol developed by Morimae and Hayashi that employs ideas from MBQC in order to perform verification [32].
2. **Subsection 3.2**: a *post-hoc* verification protocol, developed by Morimae and Fitzsimons [30] (and independently by Hangleiter et al [31]).

There is an additional receive-and-measure protocol by Gheorghiu, Wallden and Kashefi [34] which we refer to as *Steering-based VUBQC*. That protocol, however, is similar to the entanglement-based GKW protocol from Subsection 4.1.2. We will therefore review Steering-based VUBQC in that subsection by comparing it to the entanglement-based protocol.

3.1. Measurement-only Verification

In this section we discuss the measurement-only protocol from [32], which we shall simply refer to as the *measurement-only protocol*. This protocol uses MBQC to perform the quantum computation, like the VUBQC protocol from Subsection 2.2, however the manner in which verification is performed is more akin to Broadbent's Test-or-Compute protocol, from Subsection 2.3. This is because, just like in the Test-or-Compute protocol, the measurement-only approach has the verifier alternate between performing the computation or testing the prover's operations.

The key idea for this protocol, is the fact that graph states can be completely specified by a set of stabilizer operators. This fact was explained in Subsection 7.2. To reiterate the main points, recall that for a graph G, with associated graph state $|G\rangle$, if we denote as $V(G)$ the set of vertices in G and as $N_G(v)$ the set of neighbours for a given vertex v, then the generators for the stabilizer group of $|G\rangle$ are:

$$K_v = \mathsf{X}_v \prod_{w \in N_G(v)} \mathsf{Z}_w \tag{50}$$

for all $v \in V(G)$. In other words, the K_v operators generate the entire group of operators, O, such that $O|G\rangle = |G\rangle$. Note that another set of generators is given by the *dual operators*:

$$J_v = \mathsf{Z}_v \prod_{w \in N_G(v)} \mathsf{X}_w \,. \tag{51}$$

When viewed as observables, stabilizers allow one to test that an unknown quantum state is in fact a particular graph state $|G\rangle$, with high probability. This is done by measuring random stabilizers of $|G\rangle$ on multiple copies of the unknown quantum state. If all measurements return the $+1$ outcome, then, the unknown state is close in trace distance to $|G\rangle$. This is related to a concept known as *self-testing*, which is the idea of determining whether an unknown quantum state and an unknown set of observables are close to a target state and observables, based on observed statistics. We postpone a further discussion of this topic to the next section, since self-testing is ubiquitous in entanglement-based protocols.

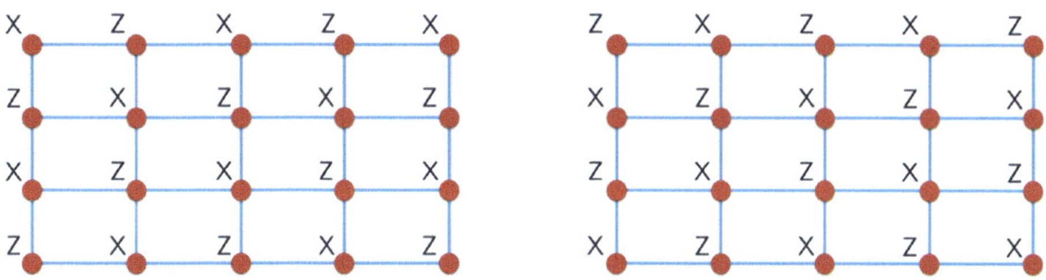

(A) XZ group measurement operators

(B) ZX group measurement operators

FIGURE 10. Stabilizer measurements

As mentioned, the measurement-only protocol involves a testing phase and a computation phase. The prover will be instructed to prepare multiple copies of a 2D cluster state, $|G\rangle$, and send them, qubit by qubit, to the verifier. The verifier will then randomly use one of these copies to perform the MBQC computation, whereas the other copies are used for testing that the correct cluster state was prepared[31]. This testing phase will involve checking all possible stabilizers of $|G\rangle$. In particular, the verifier will divide the copies to be tested into two groups, which we shall refer to as the XZ group and the ZX group. In the XZ group of states, the verifier will measure the qubits according to the 2D cluster structure, starting with an X operator in the upper left corner of the lattice and then alternating between X and Z. In the ZX group, she will measure the dual operators by swapping X with Z. The two cases are illustrated in Figure 10.

Together, the measurement outcomes of the two groups can be used to infer the outcomes of all stabilizer measurements consisting of either K_v operators or J_v

[31]This is very much in the spirit of a cryptographic technique known as *cut-and-choose* [60], which has also been used in the context of testing quantum states [61].

operators (but not combinations of both). For instance, assuming the measurement outcomes are ± 1 (the eigenvalues of the operators), to compute the outcome of a K_v or J_v stabilizer, for some node v, the verifier simply takes product of the measurement outcomes for all nodes in $\{v\} \cup N_v$. These tests allow the verifier to certify that the prover is indeed preparing copies of the state $|G\rangle$. She can then use one of these copies to run the computation. Since the prover does not know which state the verifier will use for the computation, any deviation he implements has a high chance of being detected by one of the verifier's tests. Hence, the protocol works as follows:

(1) The verifier chooses an input x and a quantum computation $\mathcal{C}$.

(2) She instructs the prover to prepare $2k+1$ copies of a 2D cluster state, $|G\rangle$, for some constant k, and send all of the qubits, one at a time, to the verifier.

(3) The verifier randomly picks one copy to run the computation of $\mathcal{C}$ on x in an MBQC fashion. The remaining $2k$ copies are randomly divided into the XZ groups and the ZX group and measured, as described above, so as to check the stabilizers of $|G\rangle$.

(4) If all stabilizer measurement outcomes are successful (i.e., produced the outcome $+1$), then the verifier accepts the outcome of the computation, otherwise she rejects.

As with all protocols, completeness follows immediately, since if the prover behaves honestly, the verifier will accept the outcome of the computation. In the case of soundness, Hayashi and Morimae treat the problem as a *hypothesis test.* In other words, in the testing phase of the protocol the verifier is checking the hypothesis that the prover prepared $2k+1$ copies of the state $|G\rangle$. Hayashi and Morimae then prove the following theorem:

Theorem 5. *Let $1/(2k+1) \leq \alpha \leq 1$ be the verifier's confidence level in the testing phase of the measurement-only protocol. Then, the state used by the verifier for the computation, denoted ρ, satisfies:*

$$\langle G|\, \rho\, |G\rangle \geq 1 - \frac{1}{\alpha(2k+1)}\,. \tag{52}$$

This theorem is essentially showing that as the number of copies of $|G\rangle$, requested by the verifier, increases, and the verifier accepts in the testing phase, one gets that the state ρ, used by the verifier for the computation, is close in trace distance to the ideal state, $|G\rangle$. The confidence level, α, represents the maximum acceptance probability for the verifier, such that the computation state, ρ, *does not* satisfy Equation 52. Essentially this represents the probability for the verifier to accept a computation state that is far from ideal. Hayashi and Morimae argue that the lower bound, $\alpha \geq 1/(2k+1)$, is tight, because if the prover corrupts one of the $2k+1$ states sent to the verifier, there is a $1/(2k+1)$ chance that that state will not be tested and the verifier accepts.

If one now denotes with C the POVM that the verifier applies on the computation state in order to perform the computation of $\mathcal{C}$, then it is the case that:

$$|Tr(C\rho) - Tr(C\,|G\rangle\,\langle G|)| \leq \frac{1}{\sqrt{\alpha(2k+1)}}. \tag{53}$$

What this means is that the distribution of measurement outcomes for the state ρ, sent by the prover in the computation run, is almost indistinguishable from the distribution of measurement outcomes for the ideal state $|G\rangle$. The soundness of the protocol is therefore upper bounded by $\frac{1}{\sqrt{\alpha(2k+1)}}$. This implies that to achieve soundness below ϵ, for some $\epsilon > 0$, the number of copies that the prover would have to prepare scales as $O\left(\frac{1}{\alpha} \cdot \frac{1}{\epsilon^2}\right)$.

In terms of the quantum capabilities of the verifier, she only requires a single qubit measurement device capable of measuring the observables: $\mathsf{X}, \mathsf{Y}, \mathsf{Z}, (\mathsf{X}+\mathsf{Y})/\sqrt{2}, (\mathsf{X}-\mathsf{Y})/\sqrt{2}$. Recently, however, Morimae, Takeuchi and Hayashi have proposed a similar protocol which uses *hypergraph states* [33]. These states have the property that one can perform universal quantum computations by measuring only the Pauli observables (X, Y and Z). Hypergraph states are generalizations of graph states in which the vertices of the graph are linked by hyperedges, which can connect more than two vertices. Hence, the entangling of qubits is done with a generalized CZ operation involving multiple qubits. The protocol itself is similar to the one from [32], as the prover is required to prepare many copies of a hypergraph state and send them to the verifier. The verifier will then test all but one of these states using stabilizer measurements and use the remaining one to perform the MBQC computation. For a computation, $\mathcal{C}$, the protocol has completeness lower bounded by $1 - |\mathcal{C}|e^{-|\mathcal{C}|}$ and soundness upper bounded by $1/\sqrt{|\mathcal{C}|}$. The communication complexity is higher than the previous measurement-only protocol, as the prover needs to send $O(|\mathcal{C}|^{21})$ copies of the $O(|\mathcal{C}|)$-qubit graph state, leading to a total communication cost of $O(|\mathcal{C}|^{22})$. We end with the following result:

Theorem 6. *The measurement-only protocols are receive-and-measure* QPIP *protocols having an inverse polynomial gap between completeness and soundness.*

3.2. Post-hoc Verification

The protocols we have reviewed so far have all been based on cryptographic primitives. There were reasons to believe, in fact, that any quantum verification protocol would have to use some form of encryption or hiding. This is due to the parallels between verification and authentication, which were outlined in Section 2. However, it was shown that this is not the case when Morimae and Fitzsimons, and independently Hangleiter et al, proposed a protocol for *post-hoc* quantum verification [30, 31]. The name "post-hoc" refers to the fact that the protocol is not interactive, requiring a single round of back and forth communication between the prover and the verifier. Moreover, verification is performed after the computation has been carried out. It should be mentioned that the first post-hoc protocol was proposed in [23], by Fitzsimons and Hajdušek, however, that protocol utilizes multiple quantum provers, and we review it in Subsection 4.3.

In this section, we will present the post-hoc verification approach, refered to as *1S-Post-hoc*, from the perspective of the Morimae and Fitzsimons paper [30]. The reason for choosing their approach, over the Hangleiter et al one, is that the entanglement-based post-hoc protocols, from Subsection 4.3, are also described using similar terminology to the Morimae and Fitzsimons paper. The protocol of Hangleiter et al is essentially identical to the Morimae and Fitzsimons one, except it is presented from the perspective of certifying the ground state of a gapped, local Hamiltonian. Their certification procedure is then used to devise a verification protocol for a class of quantum simulation experiments, with the purpose of demonstrating a quantum computational advantage [31].

The starting point is the complexity class QMA, for which we have stated the definition in Subsection 7.3. Recall, that one can think of QMA as the class of problems for which the solution can be checked by a BQP verifier receiving a quantum state $|\psi\rangle$, known as a witness, from a prover. We also stated the definition of the k-local Hamiltonian problem, a complete problem for the class QMA, in Definition 9. We mentioned that for $k = 2$ the problem is QMA-complete. For the post-hoc protocol, Morimae and Fitzsimons consider a particular type of 2-local Hamiltonian known as an XZ-Hamiltonian.

To define an XZ-Hamiltonian we introduce some helpful notation. Consider an n-qubit operator S, which we shall refer to as XZ-term, such that $S = \bigotimes_{j=1}^{n} P_j$, with $P_j \in \{I, \mathsf{X}, \mathsf{Z}\}$. Denote $w_X(S)$ as the X-weight of S, representing the total number of j's for which $P_j = \mathsf{X}$. Similarly denote $w_Z(S)$ as the Z-weight for S. An XZ-Hamiltonian is then a 2-local Hamiltonian of the form $H = \sum_i a_i S_i$, where the a_i's are real numbers and the S_i's are XZ-terms having $w_X(S_i) \leq 1$ and $w_Z(S_i) \leq 1$. Essentially, as the name suggests, an XZ-Hamiltonian is one in which each local term has at most one X operator and one Z operator.

The 1S-Post-hoc protocol starts with the observation that $\mathsf{BQP} \subseteq \mathsf{QMA}$. This means that any problem in BQP can be viewed as an instance of the 2-local Hamiltonian problem. Therefore, for any language $L \in \mathsf{BQP}$ and input x, there exists an XZ-Hamiltonian, H, such that the smallest eigenvalue of H is less than a when $x \in L$ or larger than b, when $x \notin L$, where a and b are a pair of numbers satisfying $b - a \geq 1/poly(|x|)$. Hence, the lowest energy eigenstate of H (also referred to as *ground state*), denoted $|\psi\rangle$, is a quantum witness for $x \in L$. In a QMA protocol, the prover would be instructed to send this state to the verifier. The verifier then performs a measurement on $|\psi\rangle$ to estimate its energy, accepting if the estimate is below a and rejecting otherwise. However, we are interested in a verification protocol for BQP problems where the verifier has minimal quantum capabilities. This means that there will be two requirements: the verifier can only perform single-qubit measurements; the prover is restricted to BQP computations. The 1S-Post-hoc protocol satisfies both of these constraints.

The first requirement is satisfied because estimating the energy of a quantum state, $|\psi\rangle$, with respect to an XZ-Hamiltonian H, can be done by measuring one of the observables S_i on the state $|\psi\rangle$. Specifically, it is shown in [62] that if one chooses the local term S_i according to a probability distribution given by the normalized terms $|a_i|$, and measures $|\psi\rangle$ with the S_i observables, this provides an estimate for the energy of $|\psi\rangle$. Since H is an XZ-Hamiltonian, this entails performing at most one X measurement and one Z measurement. This implies that the verifier need only perform single-qubit measurements.

For the second requirement, one needs to show that for any BQP computation, there exists an XZ-Hamiltonian such that the ground state can be prepared by a polynomial-size quantum circuit. Suppose the computation that the verifier would like to delegate is denoted as $\mathcal{C}$ and the input for this computation is x. Given what we have mentioned above, regarding the local Hamiltonian problem, it follows that there exists an XZ-Hamiltonian H and numbers a and b, with $b - a \geq 1/poly(|x|)$, such that if $\mathcal{C}$ accepts x with high probability then the ground state of H has energy below a, otherwise it has energy above b. It was shown in [63, 64], that starting from $\mathcal{C}$ and x one can construct an XZ-Hamiltonian satisfying this property and which also has a ground state that can be prepared by a BQP machine. The ground state is known as the *Feynman-Kitaev clock state*. To describe this state, suppose the circuit $\mathcal{C}$ has T gates (i.e., $T = |\mathcal{C}|$) and that these gates, labelled in the order in which they are applied, are denoted $\{U_i\}_{i=0}^{T}$. For $i = 0$ we assume $U_0 = I$. The Feynman-Kitaev state is the following:

$$|\psi\rangle = \frac{1}{\sqrt{T+1}} \sum_{t=0}^{T+1} \prod_{i=0}^{t} U_i \, |x\rangle \, |1^t 0^{T-t}\rangle \, . \tag{54}$$

This is essentially a superposition over all time steps of the time evolved state in the circuit $\mathcal{C}$. Hence, the state can be prepared by a BQP machine. The XZ-Hamiltonian, proposed by Kempe, Kitaev and Regev [64], is then a series of 2-local constraints that are all simultaneously satisfied by this state.

We can now present the steps of the 1S-Post-hoc protocol:

(1) The verifier chooses a quantum circuit, $\mathcal{C}$, and an input x to delegate to the prover.

(2) The verifier computes the terms a_i of the XZ-Hamiltonian, $H = \sum_i a_i S_i$, having as a ground state the Feynman-Kitaev state associated with $\mathcal{C}$ and x, denoted $|\psi\rangle$.

(3) The verifier instructs the prover to send her $|\psi\rangle$, qubit by qubit.

(4) The verifier chooses one of the XZ-terms S_i, according to the normalized distribution $\{|a_i|\}_i$, and measures it on $|\psi\rangle$. She accepts if the measurement indicates the energy of $|\psi\rangle$ is below a.

Note that the protocol is not blind, since the verifier informs the prover about both the computation $\mathcal{C}$ and the input x.

As mentioned, the essential properties that any QPIP protocol should satisfy are completeness and soundness. For the post-hoc protocol, these follow immediately from the local Hamiltonian problem. Specifically, we know that there exist a

and b such that $b - a \geq 1/poly(|x|)$. When $\mathcal{C}$ accepts x with high probability, the state $|\psi\rangle$ will be an eigenstate of H having eigenvalue smaller than a. Otherwise, any state, when measured under the H observable, will have an energy greater than b. Of course, the verifier is not computing the exact energy $|\psi\rangle$ under H, merely an estimate. This is because she is measuring only one local term from H. However, it is shown in [30] that the precision of her estimate is also inverse polynomial in $|x|$. Therefore:

Theorem 7. *1S-Post-hoc is a receive-and-measure* QPIP *protocol having an inverse polynomial gap between completeness and soundness.*

The only quantum capability of the verifier is the ability to measure single qubits in the computational and Hadamard bases (i.e., measuring the Z and X observables). The protocol, as described, suggests that it is sufficient for the verifier to measure only two qubits. However, since the completeness-soundness gap decreases with the size of the input, in practice one would perform a sequential repetition of this protocol in order to boost this gap. It is easy to see that, for a protocol with a completeness-soundness gap of $1/p(|x|)$, for some polynomial p, in order to achieve a constant gap of at least $1 - \epsilon$, where $\epsilon > 0$, the protocol needs to be repeated $O(p(|x|) \cdot \log(1/\epsilon))$ times. It is shown in [31, 65] that $p(|x|)$ is $O(|\mathcal{C}|^2)$, hence the protocol should be repeated $O(|\mathcal{C}|^2 \cdot \log(1/\epsilon))$ times and this also gives us the total number of measurements for the verifier[32]. Note, however, that this assumes that each run of the protocol is independent of the previous one (in other words, that the states sent by the prover to the verifier in each run are uncorrelated). Therefore, the $O(|\mathcal{C}|^2 \cdot \log(1/\epsilon))$ overhead should be taken as an i.i.d. (independent and identically distributed states) estimate. This is, in fact, mentioned explicitly in the Hangleiter et al result, where they explain that the prover should prepare "*a number of independent and identical copies of a quantum state*" [31]. Thus, when considering the most general case of a malicious prover that does not obey the i.i.d. constraint, one requires a more thorough analysis involving non-independent runs, as is done in the measurement-only protocol [32] or the steering-based VUBQC protocol [34].

3.3. Summary of receive-and-measure protocols

Receive-and-measure protocols are quite varied in the way in which they perform verification. The measurement-only protocols use stabilizers to test that the prover prepared a correct graph state and then has the verifier use this state to perform an MBQC computation. The 1S-Post-hoc protocol relies on the entirely different approach of estimating the ground state energy of a local Hamiltonian. Lastly, the steering-based VUBQC protocol, which we detail in Subsection 4.1.2, is different from these other two approaches by having the verifier remotely prepare

[32]As a side note, the total number of measurements is not the same as the communication complexity for this protocol, since the prover would have to send $O(|\mathcal{C}|^3 \cdot \log(1/\epsilon))$ qubits in total. This is because, for each repetition, the prover sends a state of $O(|\mathcal{C}|)$ qubits, but the verifier only measures 2 qubits from each such state.

the VUBQC states on the prover's side and then doing trap-based verification. Having such varied techniques leads to significant differences in the total number of measurements performed by the verifier, as we illustrate in Table 2.

Protocol	Measurements	Observables	Blind
Measurement-only	$O(N \cdot 1/\alpha \cdot 1/\epsilon^2)$	5	Y
Hypergraph measurement-only	$O(\max(N, 1/\epsilon^2)^{22})$	3	Y
1S-Post-hoc	$O(N^2 \cdot \log(1/\epsilon))$	2	N
Steering-based VUBQC	$O(N^{13} \log(N) \cdot \log(1/\epsilon))$	5	Y

TABLE 2. Comparison of receive-and-measure protocols. We denote $N = |\mathcal{C}|$ to be the size of the delegated quantum computation. The number of measurements is computed for a target gap between completeness and soundness of $1 - \epsilon$ for some constant $\epsilon > 0$. For the first measurement-only protocol, α denotes the confidence level of the verifier in the hypothesis test.

Of course, the number of measurements is not the only metric we use in comparing the protocols. Another important aspect is how many observables the verifier should be able to measure. The 1S-Post-hoc protocol is optimal in that sense, since the verifier need only measure X and Z observables. Next is the hypergraph state measurement-only protocol which requires all three Pauli observables. Lastly, the other two protocols require the verifier to be able to measure the XY-plane observables X, Y, $(\mathsf{X} + \mathsf{Y})/\sqrt{2}$ and $(\mathsf{X} - \mathsf{Y})/\sqrt{2}$ plus the Z observable.

Finally, we compare the protocols in terms of blindness, which we have seen plays an important role in prepare-and-send protocols. For receive-and-measure protocols, the 1S-Post-hoc protocol is the only one that is not blind. While this is our first example of a verification protocol that does not hide the computation and input from the prover, it is not the only one. In the next section, we review two other post-hoc protocols that are also not blind.

4. Entanglement-based Protocols

The protocols discussed in the previous sections have been either prepare-and-send or receive-and-measure protocols. Both types employ a verifier with some minimal quantum capabilities interacting with a single BQP prover. In this section we explore protocols which utilize multiple non-communicating provers that share entanglement and a fully classical verifier. The main idea will be for the verifier to distribute a quantum computation among many provers and verify its correct execution from correlations among the responses of the provers.

We classify the entanglement-based approaches as follows:

1. **Subsection 4.1** three protocols which make use of the CHSH game, the first one developed by Reichardt, Unger and Vazirani [19], the second by Gheorghiu, Kashefi and Wallden [20] and the third by Hajdušek, Pérez-Delgado and Fitzsimons.
2. **Subsection 4.2** a protocol based on self-testing graph states, developed by McKague [22].
3. **Subsection 4.3** two post-hoc protocols, one developed by Fitzsimons and Hajdušek [30] and another by Natarajan and Vidick [24].

Unlike the previous sections where, for the most part, each protocol was based on a different underlying idea for performing verification, entanglement-based protocols are either based on some form of rigid self-testing or on testing local Hamiltonians via the post-hoc approach. In fact, as we will see, even the post-hoc approaches employ self-testing. Of course, there are distinguishing features within each of these broad categories, but due to their technical specificity, we choose to label the protocols in this section by the initials of the authors.

Since self-testing plays such a crucial role in entanglement-based protocols, let us provide a brief description of the concept. The idea of self-testing was introduced by Mayers and Yao in [66], and is concerned with characterising the shared quantum state and observables of n non-communicating players in a *non-local game*. A non-local game is one in which a referee (which we will later identify with the verifier) will ask questions to the n players (which we will identify with the provers) and, based on their responses, decide whether they win the game or not. Importantly, we are interested in games where there is a quantum strategy that outperforms a classical strategy. By a classical strategy, we mean that the players can only produce local correlations[33]. Conversely, in a quantum strategy, the players are allowed to share entanglement in order to produce non-local correlations and achieve a higher win rate. Even so, there is a limit to how well the players can perform in the game. In other words, the optimal quantum strategy has a certain probability of winning the game, which may be less than 1. Self-testing results are concerned with non-local games in which the optimal quantum strategy is *unique*, up to local isometries on the players' systems. This means that if the referee observes a near maximal win rate for the players, in the game, she can conclude that they are using the optimal strategy and can therefore characterise their shared state and their observables, up to a local isometries. More formally, we give the definition of self-testing, adapted from [67] and using notation similar to that of [24]:

Definition 4 (Self-testing). *Let G denote a game involving n non-communicating players denoted $\{P_i\}_{i=1}^n$. Each player will receive a question from a set, Q and reply*

[33]To define local correlations, consider a setting with two players, Alice and Bob. Each player receives an input, x for Alice and y for Bob and produces an output, denoted a for Alice and b for Bob. We say that the players' responses are locally correlated if: $Pr(a,b|x,y) = \sum_\lambda Pr(a|x,\lambda)Pr(b|y,\lambda)Pr(\lambda)$. Where λ is known as a *hidden variable*. In other words, given this hidden variable, the players' responses depend only on their local inputs.

with an answer from a set A. Thus, each P_i can be viewed as a mapping from Q to A. There exists some condition establishing which combinations of answers to the questions constitutes a win for the game. Let $\omega^*(G)$ denote the maximum winning probability of the game for players obeying quantum mechanics.

The mappings P_i are implemented by a measurement strategy $S = (|\psi\rangle, \{O_i^j\}_{ij})$ consisting of a state $|\psi\rangle$ shared among the n players and local observables $\{O_i^j\}_j$, for each player P_i. We say that the game G ***self-tests*** *the strategy S, with robustness $\epsilon = \epsilon(\delta)$, for some $\delta > 0$, if, for any strategy $\tilde{S} = (|\tilde{\psi}\rangle, \{\tilde{O}_i^j\}_{ij})$ achieving winning probability $\omega^*(G) - \epsilon$ there exists a local isometry $\Phi = \bigotimes_{i=1}^n \Phi_i$ and a state $|junk\rangle$ such that:*

$$TD(\Phi(|\tilde{\psi}\rangle), |junk\rangle |\psi\rangle) \leq \delta \tag{55}$$

and for all j:

$$TD\left(\Phi\left(\bigotimes_{i=1}^{n} \tilde{O}_i^j |\tilde{\psi}\rangle\right), |junk\rangle \bigotimes_{i=1}^{n} O_i^j |\psi\rangle\right) \leq \delta\,. \tag{56}$$

4.1. Verification Based on CHSH Rigidity

4.1.1. RUV protocol. In [68], Tsirelson gave an upper bound for the total amount of non-local correlations shared between two non-communicating parties, as predicted by quantum mechanics. In particular, consider a two-player game consisting of Alice and Bob. Alice is given a binary input, labelled a, and Bob is given a binary input, labelled b. They each must produce a binary output and we label Alice's output as x and Bob's output as y. Alice and Bob win the game iff $a \cdot b = x \oplus y$. The two are *not* allowed to communicate during the game, however they are allowed to share classical or quantum correlations (in the form of entangled states). This defines a non-local game known as the *CHSH game* [69]. The optimal *classical* strategy for winning the game achieves a success probability of 75%, whereas, what Tsirelson proved, is that any *quantum* strategy achieves a success probability of at most $cos^2(\pi/8) \approx 85.3\%$. This maximal winning probability, in the quantum case, can in fact be achieved by having Alice and Bob do the following. First, they will share the state $|\Phi_+\rangle = (|00\rangle + |11\rangle)/\sqrt{2}$. If Alice receives input $a = 0$, then she will measure the Pauli X observable on her half of the $|\Phi_+\rangle$ state, otherwise (when $a = 1$) she measures the Pauli Z observable. Bob, on input $b = 0$ measures $(\mathsf{X} + \mathsf{Z})/\sqrt{2}$, on his half of the Bell pair, and on input $b = 1$, he measures $(\mathsf{X} - \mathsf{Z})/\sqrt{2}$. We refer to this strategy as the *optimal quantum strategy* for the CHSH game.

McKague, Yang and Scarani proved a converse of Tsierlson's result, by showing that if one observes two players winning the CHSH game with a near $cos^2(\pi/8)$ probability, then it can be concluded that the players' shared state is close to a Bell pair and their observables are close to the ideal observables of the optimal strategy (Pauli X and Z, for Alice, and $(\mathsf{X}+\mathsf{Z})/\sqrt{2}$ and $(\mathsf{X}-\mathsf{Z})/\sqrt{2}$, for Bob) [70]. This is effectively a self-test for a Bell pair. Reichardt, Unger and Vazirani then proved a more general result for self-testing a *tensor product* of multiple Bell states

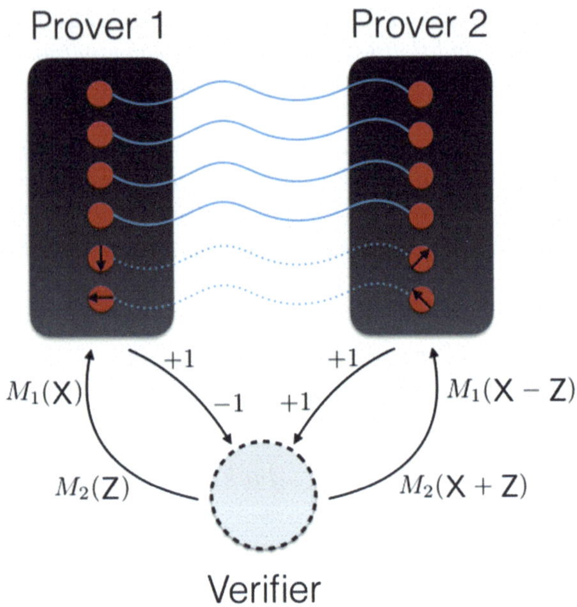

FIGURE 11. Ideal CHSH game strategy

as well as the observables acting on these states [19][34]. It is this latter result that is relevant for the RUV protocol so we give a more formal statement for it:

Theorem 8. *Suppose two players, Alice and Bob, are instructed to play n sequential CHSH games. Let the inputs, for Alice and Bob, be given by the n-bit strings* $\mathbf{a}, \mathbf{b} \in \{0,1\}^n$. *Additionally, let* $S = (|\tilde{\psi}\rangle, \tilde{A}(\mathbf{a}), \tilde{B}(\mathbf{b}))$ *be the strategy employed by Alice and Bob in playing the n CHSH games, where* $|\tilde{\psi}\rangle$ *is their shared state and* $\tilde{A}(\mathbf{a})$ *and* $\tilde{B}(\mathbf{b})$ *are their respective observables, for inputs* $\mathbf{a}, \mathbf{b}$.

Suppose Alice and Bob win at least $n(1-\epsilon)cos^2(\pi/8)$ *games, with* $\epsilon = poly(\delta, 1/n)$ *for some* $\delta > 0$, *such that* $\epsilon \to 0$ *as* $\delta \to 0$ *or* $n \to \infty$. *Then, there exist a local isometry* $\Phi = \Phi_A \otimes \Phi_B$ *and a state* $|junk\rangle$ *such that:*

$$TD(\Phi(|\tilde{\psi}\rangle), |junk\rangle |\Phi_+\rangle^{\otimes n}) \leq \delta \tag{57}$$

and:

$$TD\left(\Phi\left(\tilde{A}(\mathbf{a}) \otimes \tilde{B}(\mathbf{b}) |\tilde{\psi}\rangle\right), |junk\rangle\, A(\mathbf{a}) \otimes B(\mathbf{b}) |\Phi_+\rangle^{\otimes n}\right) \leq \delta \tag{58}$$

[34]Note that the McKague, Yang and Scarani result could also be used to certify a tensor product of Bell pairs, by repeating the self-test of a single Bell pair multiple times. However, this would require each repetition to be independent of the previous one. In other words the states shared by Alice and Bob, as well as their measurement outcomes, should be independent and identically distributed (i.i.d.) in each repetition. The Reichardt, Unger and Vazirani result makes no such assumption.

where $A(\mathbf{a}) = \bigotimes_{i=1}^{n} P(\mathbf{a}(\mathbf{i}))$, $B(\mathbf{b}) = \bigotimes_{i=1}^{n} Q(\mathbf{b}(\mathbf{i}))$ *and* $P(0) = \mathsf{X}$, $P(1) = \mathsf{Z}$, $Q(0) = (\mathsf{X} + \mathsf{Z})/\sqrt{2}$, $Q(1) = (\mathsf{X} - \mathsf{Z})/\sqrt{2}$.

What this means is that, up to a local isometry, the players share a state which is close in trace distance to a tensor product of Bell pairs and their measurements are close to the ideal measurements. This result, known as *CHSH game rigidity*, is the key idea for performing multi-prover verification using a classical verifier. We will refer to the protocol in this section as the *RUV protocol.*

Before giving the description of the protocol, let us first look at an example of *gate teleportation*, which we also mentioned when presenting the Poly-QAS VQC protocol of Subsection 2.1.2. Suppose two parties, Alice and Bob, share a Bell state $|\Phi_+\rangle$. Bob applies a unitary U on his share of the entangled state so that the joint state becomes $(I \otimes U) |\Phi_+\rangle$. Alice now takes an additional qubit, labelled $|\psi\rangle$ and measures this qubit and the one from the $|\Phi_+\rangle$ state in the Bell basis given by the states:

$$|\Phi_+\rangle = \frac{|00\rangle + |11\rangle}{\sqrt{2}}, \quad |\Phi_-\rangle = \frac{|00\rangle - |11\rangle}{\sqrt{2}},$$

$$|\Psi_+\rangle = \frac{|01\rangle + |10\rangle}{\sqrt{2}} \quad |\Psi_-\rangle = \frac{|01\rangle - |10\rangle}{\sqrt{2}}.$$

The outcome of this measurement will be two classical bits which we label b_1 and b_2. After the measurement, the state on Bob's system will be $U\mathsf{X}^{b_1}\mathsf{Z}^{b_2} |\psi\rangle$. Essentially, Bob has a one-time padded version of $|\psi\rangle$ with the U gate applied.

We now describe the RUV protocol. It uses two quantum provers but can be generalized to any number of provers greater than two. Suppose that Alice and Bob are the two provers. They are allowed to share an unbounded amount of quantum entanglement but are not allowed to communicate during the protocol. A verifier will interact classically with both of them in order to delegate and check an arbitrary quantum computation specified by the quantum circuit $\mathcal{C}$. The protocol consists in alternating randomly between four sub-protocols.

- **CHSH games.** In this subprotocol, the verifier will simply play CHSH games with Alice and Bob. To be precise, the verifier will repeatedly instruct Alice and Bob to perform the ideal measurements of the CHSH game. She will collect the answers of the two provers (which we shall refer to as CHSH statistics) and after a certain number of games, will compute the win rate of the two provers. The verifier is interested in the case when Alice and Bob win close to the maximum number of games as predicted by quantum mechanics. Thus, at the start of the protocol she takes $\epsilon = poly(1/|\mathcal{C}|)$ and accepts the statistics produced by Alice and Bob if and only if they win at least a fraction $(1 - \epsilon)cos^2(\pi/8)$ of the total number of games. Using the rigidity result, this implies that Alice and Bob share a state which is close to a tensor product of perfect Bell states (up to a local isometry). This step is schematically illustrated in Figure 11.

- **State tomography.** This time the verifier will instruct Alice to perform the ideal CHSH game measurements, as in the previous case. However, she instructs Bob to measure his halves of the entangled states so that they collapse to a set of *resource states* which will be used to perform gate teleportation. The resource states are chosen so that they are universal for quantum computation. Specifically, in the RUV protocol, the following resource states are used: $\{\mathsf{P}\left|0\right\rangle, (\mathsf{HP})_2\left|\Phi_+\right\rangle, (\mathsf{GY})_2\left|\Phi_+\right\rangle, \mathsf{CNOT}_{2,4}\mathsf{P}_2\mathsf{Q}_4(\left|\Phi_+\right\rangle \otimes \left|\Phi_+\right\rangle) : \mathsf{P}, \mathsf{Q} \in \{\mathsf{X}, \mathsf{Y}, \mathsf{Z}, I\}\}$, where $\mathsf{G} = exp\left(-i\frac{\pi}{8}\mathsf{Y}\right)$ and the subscripts indicate on which qubits the operators act. Assuming Alice and Bob do indeed share Bell states, Bob's measurements will collapse Alice's states to the same resource states (up to a one-time padding known to the verifier). Alice's measurements on these states are used to check Bob's preparation, effectively performing state tomography on the resource states.
- **Process tomography.** This subprotocol is similar to the state tomography one, except the roles of Alice and Bob are reversed. The verifier instructs Bob to perform the ideal CHSH game measurements. Alice, on the other hand, is instructed to perform Bell basis measurements on pairs of qubits. As in the previous subprotocol, Bob's measurement outcomes are used to tomographically check that Alice is indeed performing the correct measurements.
- **Computation.** The final subprotocol combines the previous two. Bob is asked to perform the resource preparation measurements, while Alice is asked to perform Bell basis measurements. This effectively makes Alice perform the desired computation through repeated gate teleportation.

An important aspect, in proving the correctness of the protocol, is the local similarity of pairs of subprotocols. For instance, Alice cannot distinguish between the CHSH subprotocol and the state tomography one, or between the process tomography one and computation. This is because, in those situations, she is asked to perform the same operations on her side, while being unaware of what Bob is doing. Moreover, since the verifier can test all but the computation part, if Alice deviates there will be a high probability of her deviation being detected. The same is true for Bob. In this way, the verifier can, essentially, enforce that the two players behave honestly and thus perform the correct quantum computation. Note, that this is not the same as the blindness property, discussed in relation to the previous protocols. The RUV protocol does, however, posses that property as well. This follows from a more involved argument regarding the way in which the computation by teleportation is performed.

It should be noted that there are only two constraints imposed on the provers: that they cannot communicate once the protocol has commenced and that they produce close to quantum optimal win-rates for the CHSH games. Importantly, there are no constraints on the quantum systems possessed by the provers, which can be arbitrarily large. Similarly, there are no constraints on what measurements they perform or what strategy they use in order to respond to the verifier. In spite of this, the rigidity result shows that for the provers to produce statistics that are accepted by the verifier, they must behave according to the ideal strategy (up to local isometry). Having the ability to fully characterise the prover's shared state and their strategies in this way is what allows the verifier to check the correctness of the delegated quantum computation. This approach, of giving a full characterisation of the states and observables of the provers, is a powerful technique which is employed by all the other entanglement-based protocols, as we will see.

In terms of practically implementing such a protocol, there are two main considerations: the amount of communication required between the verfier and the provers and the required quantum capabilities of the provers. For the latter, it is easy to see that the RUV protocol requires both provers to be universal quantum computers (i.e., BQP machines), having the ability to store multiple quantum states and perform quantum circuits on these states. In terms of the communication complexity, since the verifier is restricted to BPP, the amount of communication must scale polynomially with the size of the delegated computation. It was computed in [20], that this communication complexity is of the order $O(|\mathcal{C}|^c)$, with $c > 8192$. Without even considering the constant factors involved, this scaling is far too large for any sort of practical implementation in the near future[35].

There are essentially two reasons for the large exponent in the scaling of the communication complexity. The first, as mentioned by the authors, is that the bounds derived in the rigidity result are not tight and could possibly be improved. The second and, arguably more important reason, stems from the rigidity result itself. In Theorem 8, notice that $\epsilon = poly(\delta, 1/n)$ and $\epsilon \to 0$ as $n \to \infty$. We also know that the provers need to win a fraction $(1 - \epsilon)cos^2(\pi/8)$ of CHSH games, in order to pass the verifier's checks. Thus, the completeness-soundness gap of the protocol will be determined by ϵ. But since, for fixed δ, ϵ is essentially inverse polynomial in n, the completeness-soundness gap will also be inverse polynomial in n. Hence, one requires polynomially many repetition in order to boost the gap to constant.

We conclude with:

Theorem 9. *The RUV protocol is an* MIP* *protocol achieving an inverse polynomial gap between completeness and soundness.*

4.1.2. GKW protocol. As mentioned, in the RUV protocol the two quantum provers must be universal quantum computers. One could ask whether this is a necessity or whether there is a way to reduce one of the provers to be non-universal. In a paper by Gheorghiu, Kashefi and Wallden it was shown that the latter option is indeed possible. This leads to a protocol which we shall refer to as the *GKW protocol*. The protocol is based on the observation that one could use the state tomography subprotocol of RUV in such a way so that one prover is remotely

[35] However, with added assumptions (such as i.i.d. states and measurement statistics for the two provers), the scaling can become small enough that experimental testing is possible. A proof of concept experiment of this is realized in [71].

preparing single qubit states for the other prover. The preparing prover would then only be required to perform single qubit measurements and, hence, not need the full capabilities of a universal quantum computer. The specific single qubit states that are chosen, can be the ones used in the VUBQC protocol of Subsection 2.2. This latter prover can then be instructed to perform the VUBQC protocol with these states. Importantly, because the provers are not allowed to communicate, this would preserve the blindness requirement of VUBQC. We will refer to the preparing prover as the *sender* and the remaining prover as the *receiver*. Once again, we assume the verifier wishes to delegate to the provers the evaluation of some quantum circuit $\mathcal{C}$.

The protocol, therefore, has a two-step structure:

(1) Verified preparation. This part is akin to the state tomography subprotocol of RUV. The verifier is trying to certify the correct preparation of states $\{|+_\theta\rangle\}_\theta$ and $|0\rangle$, $|1\rangle$, where $\theta \in \{0, \pi/4, \ldots, 7\pi/4\}$. Recall that these are the states used in VUBQC. We shall refer to them as the *resource states*. This is done by self-testing a tensor product of Bell pairs and the observables of the two provers using CHSH games and the rigidity result of Theorem 8[36]. As in the RUV protocol, the verifier will play multiple CHSH games with the provers. This time, however, each game will be an extended CHSH game (as defined in [19]) in which the verifier will ask each prover to measure an observable from the set $\{\mathsf{X}, \mathsf{Y}, \mathsf{Z}, (\mathsf{X}\pm\mathsf{Z})/\sqrt{2}, (\mathsf{Y}\pm\mathsf{Z})/\sqrt{2}, (\mathsf{X}\pm\mathsf{Y})/\sqrt{2}\}$. Alternatively, this can be viewed as the verifier choosing to play one of 6 possible CHSH games defined by the observables in that set[37] These observables are sufficient for obtaining the desired resource states. In particular, measuring the X, Y, and $(\mathsf{X} \pm \mathsf{Y})/\sqrt{2}$ observables on the Bell pairs will collapse the entangled qubits to states of the form $\{|+_\theta\rangle\}_\theta$, while measuring Z will collapse them to $|0\rangle$, $|1\rangle$. The verifier accepts if the provers win a fraction $(1-\epsilon)cos^2(\pi/8)$ of the CHSH games, where $\epsilon = poly(\delta, 1/|\mathcal{C}|)$, and $\delta > 0$ is the desired trace distance between the reduced state on the receiver's side and the ideal state consisting of the required resource states in tensor product, up to a local isometry ($\epsilon \to 0$ as $\delta \to 0$ or $|\mathcal{C}| \to \infty$). The verifier will also instruct the sender prover to perform additional measurements so as to carry out the remote preparation on the receiver's side. This verified preparation is illustrated in Figure 12.

(2) Verified computation. This part involves verifying the actual quantum computation, $\mathcal{C}$. Once the resource states have been prepared on the receiver's side, the verifier will perform the VUBQC protocol with that prover as if she had sent him the resource states. She accepts the outcome of the computation if all trap measurements succeed, as in VUBQC.

[36]In fact, what is used here is a more general version of Theorem 8 involving an extended CHSH game. See the appendix section of [19].

[37]For instance, one game would involve Alice measuring either X or Y, whereas Bob should measure $(\mathsf{X}+\mathsf{Y})/\sqrt{2}$ or $(\mathsf{X}-\mathsf{Y})/\sqrt{2}$. Similar games can be defined by suitably choosing observables from the given set.

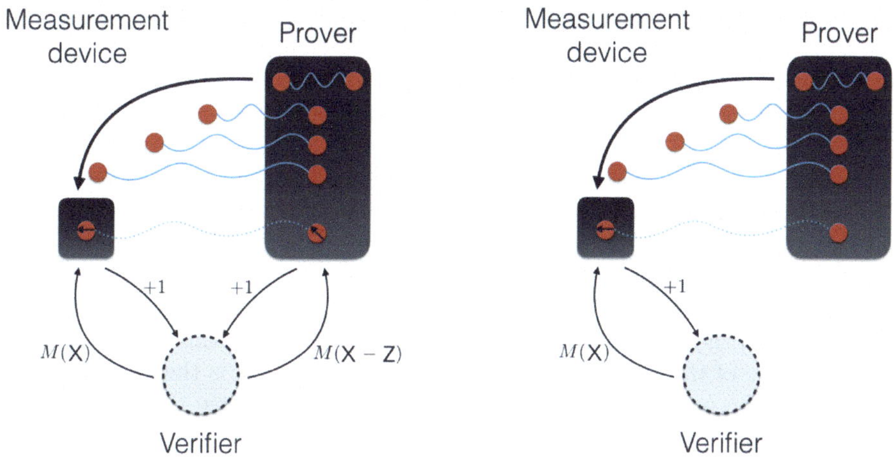

(A) Testing preparation (B) Preparing a qubit

FIGURE 12. Verified preparation

Note the essential difference, in terms of the provers' requirements, between this protocol and the RUV protocol. In the RUV protocol, both provers had to perform entangling measurements. However, in the GKW protocol, the sender prover is required to only perform single qubit measurements. This means that the sender prover can essentially be viewed as an untrusted measurement device, whereas the receiver is the only universal quantum computer. For this reason, the GKW protocol is also described as a *device-independent* [72, 73] verification protocol. This stems from comparing it to VUBQC or the receive-and-measure protocols, of Section 3, where the verifier had a trusted preparation or measurement device. In this case, the verifier essentially has a measurement device (the sender prover) which is untrusted.

Of course, performing the verified preparation subprotocol and combining it with VUBQC raises some questions. For starters, in the VUBQC protocol, the state sent to the prover is assumed to be an ideal state (i.e., an exact tensor product of states of the form $|+_\theta\rangle$ or $|0\rangle$, $|1\rangle$). However, in this case the preparation stage is probabilistic in nature and therefore the state of the receiver will be δ-close to the ideal tensor product state, for some $\delta > 0$. How is the completeness-soundness gap of the VUBQC protocol affected by this? Stated differently, is VUBQC robust to deviations in the input state? A second aspect is that, since the resource state is prepared by the untrusted sender, even though it is δ-close to ideal, it can, in principle, be correlated with the receiving prover's system. Do these initial correlations affect the security of the protocol?

Both of these issues are addressed in the proofs of the GKW protocol. Firstly, assume that in the VUBQC protocol the prover receives a state which is δ-close to ideal and uncorrelated with his private system. Any action of the prover can, in the most general sense, be modelled as a CPTP map. This CPTP map is of course distance preserving and so the output of this action will be δ-close to the output in the ideal case. It follows from this that the probabilities of the verifier accepting a correct or incorrect result change by at most $O(\delta)$. As long as $\delta > 1/poly(|\mathcal{C}|)$ (for a suitably chosen polynomial), the protocol remains a valid QPIP protocol.

Secondly, assume now that the δ-close resource state is correlated with the prover's private system, in VUBQC. It would seem that the prover could, in principle, exploit this correlation in order to convince the verifier to accept an incorrect outcome. However, it is shown that this is, in fact, not the case, as long as the correlations are small. Mathematically, let ρ_{VP} be the state comprising of the resource state and the prover's private system. In the ideal case, this state should be a product state of the form $\rho_V \otimes \rho_P$, where $\rho_V = |\psi_{id}\rangle \langle\psi_{id}|$ is the ideal resource state and ρ_P the prover's system. However, in the general case the state can be entangled. In spite of this, it is known that:

$$TD(Tr_P(\rho_{VP}), |\psi_{id}\rangle \langle\psi_{id}|) \leq \delta\,. \tag{59}$$

Using a result known as the *gentle measurement lemma* [19], one can show that this implies:

$$TD(\rho_{VP}, |\psi_{id}\rangle \langle\psi_{id}| \otimes Tr_V(\rho_{VP})) \leq O(\sqrt{\delta})\,. \tag{60}$$

In other words, the joint system of resource states and the prover's private memory is $O(\sqrt{\delta})$-close to the ideal system. Once again, as long as $\delta > 1/poly(|\mathcal{C}|)$ (for a suitably chosen polynomial), the protocol is a valid QPIP protocol.

These two facts essentially show that the GKW protocol is a valid entanglement-based protocol, as long as sufficient tests are performed in the verified preparation stage so that the system of resource states is close to the ideal resource states. As with the RUV protocol, this implies a large communication overhead, with the communication complexity being of the order $O(|\mathcal{C}|^c)$, where $c > 2048$. One therefore has:

Theorem 10. *The GKW protocol is an* MIP* *protocol achieving an inverse polynomial gap between completeness and soundness.*

Before concluding this section, we describe the steering-based VUBQC protocol that we referenced in Section 3. As mentioned, the GKW protocol can be viewed as a protocol involving a verifier with an untrusted measurement device interacting with a quantum prover. In a subsequent paper, Gheorghiu, Wallden and Kashefi addressed the setting in which the verifier's device becomes trusted [34]. They showed that one can define a self-testing game for Bell states which involves *steering correlations* [74] as opposed to non-local correlations. Steering correlations arise in a two-player setting in which one of the players is trusted to measure certain observables. This extra piece of information allows for the characterisation of Bell states with comparatively fewer statistics than in the non-local case. The steering-based VUBQC protocol, therefore, has exactly the same structure as the GKW protocol. First, the verifier uses this steering-based game, between her measurement device and the prover, to certify that the prover prepared a tensor product of Bell pairs. She then measures some of the Bell pairs so as to remotely prepare the resource states of VUBQC on the prover's side and then performs the trap-based verification. As mentioned in Section 3, the protocol has a communication complexity of $O(|\mathcal{C}|^{13}\log(|\mathcal{C}|))$ which is clearly an improvement over $O(|\mathcal{C}|^{2048})$. This improvement stems from the trust added to the measurement device. However, the overhead is still too great for any practical implementation.

4.1.3. HPDF protocol. Independently from the GKW approach, Hajdušek, Pérez-Delgado and Fitzsimons developed a protocol which also combines the CHSH rigidity result with the VUBQC protocol. This protocol, which we refer to as the *HPDF protocol* has the same structure as GKW in the sense that it is divided into a verified preparation stage and a verified computation stage. The major difference is that the number of non-communicating provers is on the order $O(poly(|\mathcal{C}|))$, where $\mathcal{C}$ is the computation that the verifier wishes to delegate. Essentially, there is one prover for each Bell pair that is used in the verified preparation stage. This differs from the previous two approaches in that the verifier knows, a priori, that there is a tensor product structure of states. She then needs to certify that these states are close, in trace distance, to Bell pairs. The advantage of assuming the existence of the tensor product structure, instead of deriving it through the RUV rigidity result, is that the overhead of the protocol is drastically reduced. Specifically, the total number of provers, and hence the total communication complexity of the protocol is of the order $O(|\mathcal{C}|^4\log(|\mathcal{C}|))$.

We now state the steps of the HPDF protocol. We will refer to one of the provers as the verifier's untrusted measurement device. This is akin to the sender prover in the GKW protocol. The remaining provers are the ones which will "receive" the states prepared by the verifier and subsequently perform the quantum computation.

(1) Verified preparation. The verifier is trying to certify the correct preparation of the resource states $\{|+_\theta\rangle\}_\theta$ and $|0\rangle$, $|1\rangle$, where $\theta \in \{0, \pi/4, \ldots, 7\pi/4\}$. The verifier instructs each prover to prepare a Bell pair and send one half to her untrusted measurement device. For each received state, she will randomly measure one of the following observables $\{\mathsf{X}, \mathsf{Y}, \mathsf{Z}, (\mathsf{X}+\mathsf{Z})/\sqrt{2}, (\mathsf{Y}+\mathsf{Z})/\sqrt{2}, (\mathsf{X}+\mathsf{Y})/\sqrt{2}, (\mathsf{X}-\mathsf{Y})/\sqrt{2}\}$. Each prover is either instructed to randomly measure an observable from the set $\{\mathsf{X}, \mathsf{Y}, \mathsf{Z}\}$ or to not perform any measurement at all. The latter case corresponds to the qubits which are prepared for the computation stage. The verifier will compute correlations between the measurement outcomes of her device and the provers and accept if these correlations are above some threshold parametrized by $\epsilon - poly(\delta, 1/|\mathcal{C}|)$ ($\epsilon \to 0$ as $\delta \to 0$ or $|\mathcal{C}| \to \infty$), where $\delta > 0$ is the desired trace distance between the reduced state on the receiving provers' sides and the ideal state consisting of the required resource states in tensor product, up to a local isometry.

(2) Verified computation. Assuming the verifier accepted in the previous stage, she instructs the provers that have received the resource states to act as a single prover. The verifier then performs the VUBQC protocol with that prover as if she had sent him the resource states. She accepts the outcome of the computation if all trap measurements succeed, as in VUBQC.

In their paper, Hajdušek et al have proved that the procedure in the verified preparation stage of their protocol constitutes a self-testing procedure for Bell states. This procedure self-tests individual Bell pairs, as opposed to the CHSH rigidity theorem which self-tests a tensor product of Bell pairs. In this case, however, the tensor product structure is already given by having the $O(|\mathcal{C}|^4 \log(|\mathcal{C}|))$ non-communicating provers. The correctness of the verified computation stage follows from the robustness of the VUBQC protocol, as mentioned in the previous section. One therefore has the following:

Theorem 11. *The HPDF protocol is an* $\mathsf{MIP}^*[\mathsf{poly}]$ *protocol achieving an inverse polynomial gap between completeness and soundness.*

4.2. Verification Based on Self-testing Graph States

We saw, in the HPDF protocol, that having multiple non-communicating provers presents a certain advantage in characterising the shared state of these provers, due to the tensor product structure of the provers' Hilbert spaces. This approach not only leads to simplified proofs, but also to a reduced overhead in characterising this state, when compared to the CHSH rigidity Theorem 8, from [19].

Another approach which takes advantage of this tensor product structure is the one of McKague from [22]. In his protocol, as in HPDF, the verifier will interact with $O(poly(|\mathcal{C}|))$ provers. Specifically, there are multiple groups of $O(|\mathcal{C}|)$ provers, each group jointly sharing a graph state $|G\rangle$. In particular, each prover should hold only one qubit from $|G\rangle$. The central idea is for the verifier to instruct the provers to measure their qubits to either test that the provers are sharing the correct graph state or to perform an MBQC computation of $\mathcal{C}$. This approach is similar to the stabilizer measurement-only protocol of Subsection 3.1 and, just like in that protocol or the Test-or-Compute or RUV protocols, the verifier will randomly alternate between tests and computation.

Before giving more details about this verification procedure, we first describe the type of graph state that is used in the protocol and the properties which allow this state to be useful for verification. McKague considers $|G\rangle$ to be a triangular graph state, which is a type of universal cluster state. What this means is that the graph G on which the state is based on, is a triangular lattice (a planar graph with triangular faces). An example is shown in Figure 13. For each vertex v in G we have that:

$$X_v Z_{N(v)} |G\rangle = |G\rangle \ . \tag{61}$$

Where $N(v)$ denotes the neighbors of the vertex v. In other words, $|G\rangle$ is stabilized by the operators $S_v = X_v Z_{N(v)}$, for all vertices v. This is important for the testing part of the protocol as it means that measuring the observable S_v will always yield the outcome 1. Another important property is:

$$X_\tau Z_{N(\tau)} |G\rangle = - |G\rangle \, , \tag{62}$$

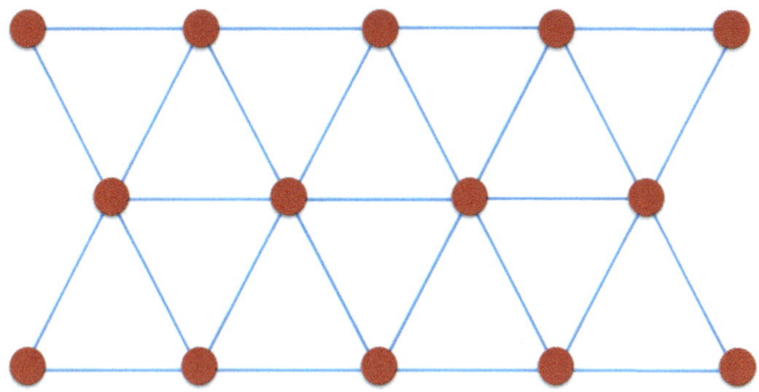

FIGURE 13. Triangular lattice graph

where τ is a set of 3 neighboring vertices which comprise a triangle in the graph G (and $N(\tau)$ are the neighbors of those vertices). This implies that measuring $T_\tau = X_\tau Z_{N(\tau)}$ produces the outcome -1. Triangular graph states are universal for quantum computation, as explained in [22], by performing local measurements (with corrections) on the vertex qubits using the observables $\mathsf{R}(\theta) = cos(\theta)\mathsf{X} + sin(\theta)\mathsf{Z}$, where $\theta \in \{0, \pi/4, \ldots, 7\pi/4\}$.

We now have the necessary elements to describe McKague's protocol. The verifier considers a triangular graph state $|G\rangle$ for the computation she wishes to verify. Let $n = O(|\mathcal{C}|)$ denote the number of vertices in G. In the ideal case, there will be multiple groups of n provers and, in each group, every prover should have one of the qubits of this graph (entangled with its neighbors). Denote T as the number of triangles (consisting of 3 neighboring vertices) in G and $N_G = 3n + T$. The protocol's setting is shown in Figure 14.

The verifier will choose one of the n groups of provers at random to perform the computation $\mathcal{C}$. The computation is performed in an MBQC fashion. In other words, the verifier will pick appropriate measurement angles $\{\theta_v\}_{v \in V(G)}$, for all vertices in G, as well as a partial order for the vertices. To perform the computation $\mathcal{C}$, the verifier instructs the provers to measure the qubits of $|G\rangle$ with the observables $\mathsf{R}(\theta_v)$, defined above. The partial order establishes the temporal ordering of these measurements. Additionally, the θ_v angles, for the $\mathsf{R}(\theta_v)$ measurements, should be updated so as to account for corrections arising from previous measurement outcomes. In other words, the angles $\{\theta_v\}_{v \in V(G)}$, which we shall refer to as *computation angles*, are the ideal angles assuming no corrections. See Subsection 7.2 for more details about measurement-based quantum computations.

The remaining groups of provers are used to test that the correct graph state, $|G\rangle$, was prepared. This testing phase consists in the verifier randomly choosing to run one of the following sub-protocols:

- **Vertex test**, chosen with probability $\frac{n}{N_G}$. The verifier randomly picks a vertex v of the graph. She instructs the prover having the qubit corresponding to v

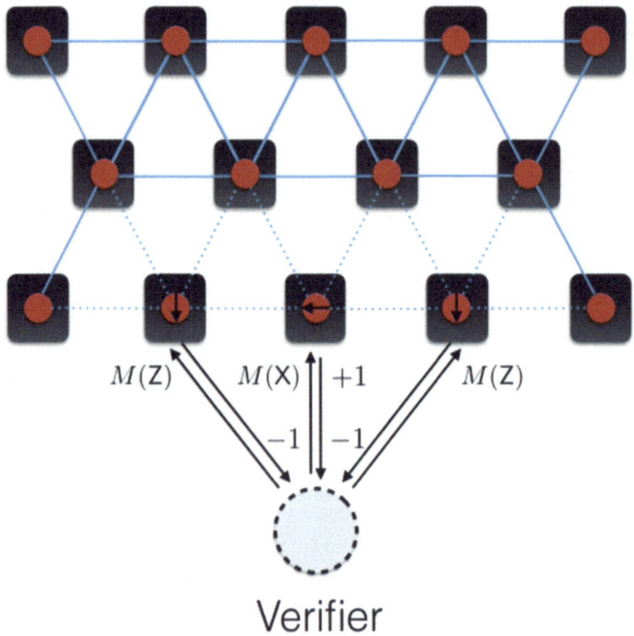

FIGURE 14. Verifier instructing some of the provers to perform measurements in McKague's protocol

to measure it in the X basis. At the same time, she asks the provers having the neighboring qubits, $N(v)$, to measure them in the Z basis. The ± 1 outcomes of the provers are multiplied by the verifier and she checks to see whether this product is 1 (since she is essentially measuring the S_v stabilizer). If this is the case, she accepts, otherwise she rejects.

- **Triangle test**, chosen with probability $\frac{T}{N_G}$. Similar to the previous test, the verifier picks a random triangle in G, consisting of the vertices τ. She instructs the provers having these vertices to measure them in the X basis, while their neighbors ($N(\tau)$) are instructed to measure in the Z basis. The verifier checks if the product of their outcomes is -1 and if so, she accepts, otherwise she rejects.
- ***R*(*θ*) test**, chosen with probability $\frac{2n}{N_G}$. In this case the verifier first randomly picks a vertex v of the graph, a neighbor u of v (so $u \in N(v)$) and t in $\{-1, +1\}$. She then randomly picks X with probability $p = \frac{cos(\theta_v)}{cos(\theta_v)+|sin(\theta_v)|}$ or Z with probability $1-p$, where θ_v is the computation angle associated with v. If she chose X, then she queries the prover holding v to measure $\mathsf{R}(t\theta_v)$, and his neighbors ($N(v)$) to measure Z. She accepts if the product of their replies

is +1. If the verifier instead chose Z, then she instructs the prover holding v to measure $t\mathsf{R}(t\theta_v)$, the prover holding u to measure X and the neighbors of u and v to measure Z. She accepts if the product of their outcomes is +1.

Together, these three tests are effectively performing a self-test of the graph state $|G\rangle$ and the prover's observables. Specifically, McKague showed the following:

Theorem 12. *For a triangular graph G, having n vertices, suppose that n provers, performing the strategy $S = (|\tilde{\psi}\rangle, \{\tilde{O}_i^j\}_{ij})$ succeed in the test described above with probability $1 - \epsilon$, where $\epsilon = poly(\delta, 1/n)$ for some $\delta > 0$ and $\epsilon \to 0$ as $\delta \to 0$ or $n \to \infty$. The strategy involves sharing the state $|\tilde{\psi}\rangle$ and measuring the observables $\{\tilde{O}_i^j\}_{ij}$, where each prover, i, has observables $\{\tilde{O}_i^j\}_j$. Then there exists a local isometry $\Phi = \bigotimes_{i=1}^n \Phi_i$ and a state $|junk\rangle$ such that:*

$$TD(\Phi(|\tilde{\psi}\rangle), |junk\rangle |G\rangle) \leq \delta \tag{63}$$

and for all j:

$$TD\left(\Phi\left(\bigotimes_{i=1}^{n} \tilde{O}_i^j |\tilde{\psi}\rangle\right), |junk\rangle \bigotimes_{i=1}^{n} O_i^j |G\rangle\right) \leq \delta \tag{64}$$

where for all i, $O_i^j \in \{\mathsf{R}(\theta) |\ \theta \in \{0, \pi/4, \ldots, 7\pi/4\}\}$[38].

Note that the verifier will ask the provers to perform the same types of measurements in both the testing phase and the computation phase of the protocol. This means that, at the level of each prover, the testing and computation phases are indistinguishable. Moreover, the triangular state $|G\rangle$, being a universal cluster state, will be the same for computations of the same size. Therefore, the protocol is blind in the sense that each prover, on its own, is unaware of what computation is being performed. In summary, the protocol consists of the verifier choosing to perform one of the following:

- **Computation.** In this case, the verifier instructs the provers to perform the MBQC computation of $\mathcal{C}$ on the graph state $|G\rangle$, as described above.
- **Testing $|G\rangle$.** In this case, the verifier will randomly choose between one of the three tests described above accepting if an only if the test succeeds.

It is therefore the case that:

Theorem 13. *McKague's protocol is an* MIP^**[poly] protocol having an inverse polynomial gap between completeness and soundness.*

As with the previous approaches, the reason for the inverse polynomial gap between completeness and soundness is the use of a self-test with robustness $\epsilon - poly(1/n)$ (and $\epsilon \to 0$ as $n \to \infty$). In turn, this leads to a polynomial overhead for the protocol as a whole. Specifically, McKague showed that the total number of required provers and communication complexity, for a quantum computation $\mathcal{C}$, is

[38] The measurement angles need not be restricted to this set, however, as in VUBQC, this set of angles is sufficient for performing universal MBQC computations.

of the order $O(|\mathcal{C}|^{22})$. Note, however, that each of the provers must only perform a single-qubit measurement. Hence, apart from the initial preparation of the graph state $|G\rangle$, the individual provers are not universal quantum computers, merely single-qubit measurement devices.

4.3. Post-hoc Verification

In Subsection 3.2 we reviewed a protocol by Morimae and Fitzsimons for post-hoc verification of quantum computation. Of course, that protocol involved a single quantum prover and a verifier with a measurement device. In this section, we review two post-hoc protocols for the multi-prover setting having a classical verifier. We start with the first post-hoc protocol by Fitzsimons and Hajdušek.

4.3.1. FH protocol. Similar to the 1S-Post-hoc protocol from Subsection 3.2, the protocol of Fitzsimons and Hajdušek, which we shall refer to as the *FH protocol*, also makes use of the local Hamiltonian problem stated in Definition 9. As mentioned, this problem is complete for the class QMA, which consists of problems that can be decided by a BQP verifier receiving a witness state from a prover. Importantly, the size of the witness state is *polynomial* in the size of the input to the problem. However, Fitzsimons and Vidick proposed a protocol for the k-local Hamiltonian problem (and hence any QMA problem), involving 5 provers, in which the quantum state received by the verifier is of *constant* size [75]. That protocol is the basis for the FH protocol and so we start with a description of it.

Suppose that the k-local Hamiltonian is of the form $H = \sum_i H_i$, acting on a system of n qubits and each H_i is a k-local, n-qubit projector. For fixed a and b, such that $b-a \geq 1/poly(n)$, the verifier should accept if there exists a state $|\psi\rangle$ such that $\langle\psi| H |\psi\rangle \leq a$ and reject if for all states $|\psi\rangle$ it is the case that $\langle\psi| H |\psi\rangle \geq b$. Suppose we are in the acceptance case and let $|\psi\rangle$ be the witness state. The 5 provers must share a version of $|\psi\rangle$ encoded in the 5-qubit error correcting code[39], denoted $|\psi\rangle_L$. Specifically, for each logical qubit of $|\psi\rangle_L$, each prover should hold one of its constituent physical qubits. The verifier will then check that the provers are indeed sharing this state, accepting if this is the case and rejecting otherwise. She will also perform an energy measurement on the state, to estimate if it has energy above b or below a. To do this she will, with equal probability, choose to either test that the shared state of the provers has energy below a or that the provers share a state encoded in the 5-qubit code:

- **Energy measurement.** In this case, the verifier will pick a random term H_i, from H, and ask each prover for k qubits corresponding to the logical states on which H_i acts. The verifier will then perform a two-outcome measurement, defined by the operators $\{H_i, I - H_i\}$ on the received qubits. As in the 1S-Post-hoc protocol, this provides an estimate for the energy of $|\psi\rangle$. The verifier accepts if the measurement outcome indicates the state has energy below a.

[39]The 5-qubit code is the smallest error correcting capable of correcting for arbitrary single-qubit errors [76].

- **Encoding measurement.** In this case the verifier will choose at random between two subtests. In the first subtest, she will choose j at random from 1 to n and ask each prover to return the physical qubits comprising the j'th logical qubit. She then measures these qubits to check whether their joint state lies within the code space, accepting if it does and rejecting otherwise. In the second subtest, the verifier chooses a random set, S, of 3 values between 1 and n. She also picks one of the values at random, labelled j. The verifier then asks a randomly chosen prover for the physical qubits of the logical states indexed by the values in S, while asking the remaining provers for their shares of logical qubit j. As an example, if the set contains the values $\{1, 5, 8\}$, then the verifier picks one of the 5 provers at random and asks him for his shares (physical qubits) of logical qubits 1, 5 and 8 from $|\psi\rangle$. Assuming that the verifier also picked the random value 8 from the set, then she will ask the remaining provers for their shares of logical qubit 8. The verifier then measures logical qubit j (or 8, in our example) and checks if it is in the code subspace, accepting if it is and rejecting otherwise. The purpose of this second subtest is to guarantee that the provers respond with different qubits when queried.

One can see that when the witness state exists and the provers follow the protocol, the verifier will indeed accept with high probability. On the other hand, Fitzsimons and Vidick show that when there is no witness state, the provers will fail at convincing the verifier to accept with high probability. This is because they cannot simultaneously provide qubits yielding the correct energy measurements and also have their joint state be in the correct code space. This also illustrates why their protocol required testing both of these conditions. If one wanted to simplify the protocol, so as to have a single prover providing the qubits for the verifier's $\{H_i, I - H_i\}$ measurement, then it is no longer possible to prove soundness. The reason is that even if there does not exist a $|\psi\rangle$ having energy less than a for H, the prover could still find a group of k qubits which minimize the energy constraint for the specific H_i that the verifier wishes to measure. The second subtest prevents this from happening, with high probability, since it forces the provers to consistently provide the requested indexed qubits from the state $|\psi\rangle$.

Note that for a BQP computation, defined by the quantum circuit $\mathcal{C}$ and input x, the state $|\psi\rangle$ in the Fitzsimons and Vidick protocol becomes the Feynman-Kitaev state of that circuit, as described in Subsection 3.2. The FH protocol essentially takes the Fitzsimons and Vidick protocol for BQP computations and alters it by making the verifier classical. This is achieved using an approach of Ji [77] which allows for the two tests to be performed using only classical interaction with the provers. The idea is based on self-testing and is similar to the rigidity of the CHSH game. To understand this approach let us first examine the stabilizer generators, $\{g_i\}_{i=1}^4$ for the code space of the 5-qubit code, shown in Table 3. Notice that they all involve only Pauli X, Z or identity operators. In particular, the operator acting on the fifth qubit is always either X or Z. Ji then considers rotating this operator

Generator	Name
IXZZX	g_1
XIXZZ	g_2
ZXIXZ	g_3
ZZXIX	g_4

TABLE 3. Generators for 5-qubit code

Generator	Name
IXZZX′	g'_1
XIXZZ′	g'_2
ZXIXZ′	g'_3
ZZXIX′	g'_4

TABLE 4. Generators with fifth operator rotated

so that $\mathsf{X} \to \mathsf{X}'$ and $\mathsf{Z} \to \mathsf{Z}'$, where $\mathsf{X}' = (\mathsf{X}+\mathsf{Z})/\sqrt{2}$ and $\mathsf{Z}' = (\mathsf{X}-\mathsf{Z})/\sqrt{2}$, resulting in the new operators $\{g'_i\}_{i=1}^4$ shown in Table 4.

The new operators satisfy a useful property. For any state $|\phi\rangle$ in the code space of the 5-qubit code, it is the case that:

$$\sum_i \langle\phi| \, g'_i \, |\phi\rangle = 4\sqrt{2}\,. \tag{65}$$

This is similar to the CHSH game. In the CHSH game, the ideal strategy involves Alice measuring either X or Z and Bob measuring either X′ or Z′, respectively, on the maximally entangled state $|\Phi_+\rangle$. These observables and the Bell state satisfy:

$$\langle\Phi_+| \, \mathsf{XX}' + \mathsf{XZ}' + \mathsf{ZX}' - \mathsf{ZZ}' \, |\Phi_+\rangle = 2\sqrt{2}\,. \tag{66}$$

It can be shown that having observables which satisfy this relation implies that Alice and Bob win the CHSH game with the (quantum) optimal probability of success $cos^2(\pi/8)$. Analogous to the CHSH game, the stabilizers $\{g'_i\}_{i=1}^4$, viewed as observables, can be used to define a 5-player non-local game, in which the optimal quantum strategy involves measuring these observables on a state encoded in the 5-qubit code. Moreover, just like in the CHSH game, observing the players achieve the maximum quantum win-rate for the game implies that, up to local isometry, the players are following the ideal quantum strategy. We will not detail the game, except to say that it inolves partitioning the 5 provers into two sets, one consisting of four provers and the other with the remaining prover. Such a bipartition of a state encoded in the 5-qubit code yields a state which is isometric to a Bell pair. This means that the 5-player game is essentially self-testing a maximally entangled state, hence the similarity to the CHSH game. This then allows a classical verifier, interacting with the 5 provers, to perform the encoding test of the Fitzsimons and Vidick protocol.

We have discussed how a classical verifier can test that the 5 provers share a state encoded in the logical space of the 5-qubit code. But to achieve the functionality of the Fitzsimons and Vidick protocol, one needs to also delegate to the provers the measurement of a local term H_i from the Hamiltonian. This is again possible using the 5-player non-local game. Firstly, it can be shown that, without loss of generality, that each H_i, in the k-local Hamiltonian, can be expressed as a linear combination of terms comprised entirely of I, X and Z. This means that the Hamiltonian itself is a linear combination of such terms, $H = \sum_i a_i S_i$, where a_i are real coefficients and S_i are k-local XZ-terms. This is akin to the XZ-Hamiltonian from the 1S-Post-hoc protocol.

Given this fact, the verifier can measure one of the S_i terms, in order to estimate the energy of the ground state, instead of measuring $\{H_i, I - H_i\}$. She will pick an S_i term at random and ask the provers to measure the constituent Pauli observables in S_i. However, the verifier will also alternate these measurements with the stabilizer measurements of the non-local game, rejecting if the provers do not achieve the maximal non-local value of the game. This essentially forces the provers to perform the correct measurements.

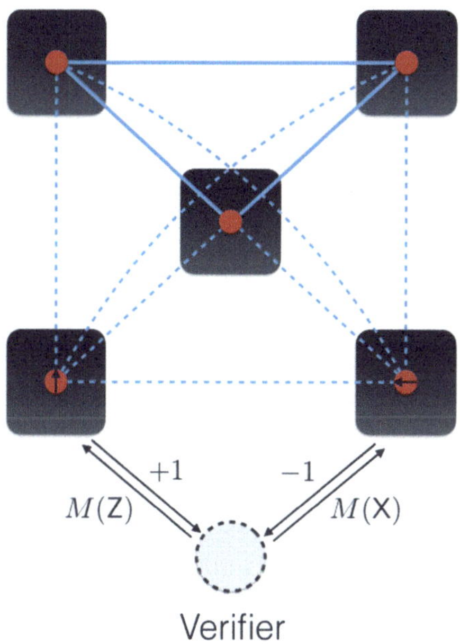

FIGURE 15. Verifier interacting with the 5 provers

To summarize, the FH protocol is a version of the Fitzsimons and Vidick protocol which restricts the provers to be BQP machines and uses Ji's techniques, based on non-local games, to make the verifier classical. The steps of the FH protocol are as follows:

(1) The verifier instructs the provers to share the Feynman-Kitaev state, associated with her circuit $\mathcal{C}$, encoded in the 5-qubit error correcting code, as described above. We denote this state as $|\psi\rangle_L$. The provers are then split up and not allowed to communicate. The verifier then considers a k-local

Hamiltonian having $|\psi\rangle_L$ as a ground state as well as the threshold values a and b, with $b - a > 1/poly(|\mathcal{C}|)$.

(2) The verifier chooses to either perform the energy measurement or the encoding measurement as described above. For the energy measurement she asks the provers to measure a randomly chosen XZ-term from the local Hamiltonian. The verifier accepts if the outcome indicates that the energy of $|\psi\rangle_L$ is below a. For the encoding measurement the verifier instructs the provers to perform the measurements of the 5-player non-local game. She accepts if the provers win the game, indicating that their shared state is correctly encoded.

One therefore has:

Theorem 14. *The FH protocol is an* MIP^* *protocol achieving an inverse polynomial gap between completeness and soundness.*

There are two significant differences between this protocol and the previous entanglement-based approaches. The first is that the protocol does not use self-testing to enforce that the provers are performing the correct operations in order to implement the computation $\mathcal{C}$. Instead, the computation is checked indirectly by using the self-testing result to estimate the ground-state energy of the k-local Hamiltonian. This then provides an answer to the considered BQP computation viewed as a decision problem[40]. The second difference is that the protocol is not blind. In all the previous approaches, the provers had to share an entangled state which was independent of the computation, up to a certain size. However, in the FH protocol, the state that the provers need to share depends on which quantum computation the verifier wishes to perform.

In terms of communication complexity, the protocol, as described, would involve only 2 rounds of interaction between the verifier and the provers. However, since the completeness-soundness gap is inverse polynomial, and therefore decreases with the size of the computation, it becomes necessary to repeat the protocol multiple times to properly differentiate between the accepting and rejecting cases. On the one hand, the local Hamiltonian itself has an inverse polynomial gap between the two cases of acceptance and rejection. As shown in [31, 65], for the Hamiltonian resulting from a quantum circuit, $\mathcal{C}$, that gap is $1/|\mathcal{C}|^2$. To boost this gap to constant, the provers must share $O(|\mathcal{C}|^2)$ copies of the Feynman-Kitaev state.

On the other hand, the self-testing result has an inverse polynomial robustness. This means that estimating the energy of the ground state is done with a precision which scales inverse polynomially in the number of qubits of the state. More precisely, according to Ji's result, the scaling should be $1/O(N^{16})$, where N is the number of qubits on which the Hamiltonian acts [77]. This means that the protocol should be repeated on the order of $O(N^{16})$ times, in order to boost the completeness-soundness gap to constant.

[40] In their paper, Fitzsimons and Hajdušek also explain how their protocol can be used to sample from a quantum circuit, rather than solve a decision problem [23].

4.3.2. NV protocol. The second entanglement-based post-hoc protocol was developed by Natarajan and Vidick [24] and we therefore refer to it as the *NV protocol.* The main ideas of the protocol are similar to those of the FH protocol. However, Natarajan and Vidick prove a self-testing result having *constant* robustness and use it in order to perform the energy estimation of the ground state for the local Hamiltonian.

The statement of their general self-testing result is too involved to state here, so instead we reproduce a corollary to their result (also from [24]) that is used for the NV protocol. This corollary involves self-testing a tensor product of Bell pairs:

Theorem 15. *For any integer n there exists a two-player non-local game, known as the Pauli braiding test (PBT), with $O(n)$-bit questions and $O(1)$-bit answers satisfying the following:*

Let $S = (|\tilde{\psi}\rangle, \tilde{A}(\mathbf{a}), \tilde{B}(\mathbf{b}))$ be the strategy employed by two players (Alice and Bob) in playing the game, where $|\tilde{\psi}\rangle$ is their shared state and $\tilde{A}(\mathbf{a})$ and $\tilde{B}(\mathbf{b})$ are their respective (multi-qubit) observables when given n-bit questions $\mathbf{a}$ and $\mathbf{b}$, respectively. Suppose Alice and Bob win the Pauli braiding test with probability $\omega^*(PBT) - \epsilon$, for some $\epsilon > 0$ (note that $\omega^*(PBT) = 1$). Then there exist $\delta = poly(\epsilon)$, a local isometry $\Phi = \Phi_A \otimes \Phi_B$ and a state $|junk\rangle$ such that:

$$TD(\Phi(|\tilde{\psi}\rangle), |junk\rangle |\Phi_+\rangle^{\otimes n}) \leq \delta \tag{67}$$

$$TD\left(\Phi\left(\tilde{A}(\mathbf{a}) \otimes \tilde{B}(\mathbf{b}) |\tilde{\psi}\rangle\right), |junk\rangle \mathsf{X}(\mathbf{a}) \otimes \mathsf{Z}(\mathbf{b}) |\Phi_+\rangle^{\otimes n}\right) \leq \delta, \tag{68}$$

where $\mathsf{X}(\mathbf{a}) = \bigotimes_{i=1}^{n} \mathsf{X}^{\mathbf{a(i)}}$ *and* $\mathsf{Z}(\mathbf{b}) = \bigotimes_{i=1}^{n} \mathsf{Z}^{\mathbf{b(i)}}$.

This theorem is essentially a self-testing result for a tensor product of Bell states, and Pauli X and Z observables, achieving a constant robustness. The Pauli braiding test is used in the NV protocol in a similar fashion to Ji's result, from the previous subsection, in order to certify that a set of provers are sharing a state that is encoded in a quantum error correcting code. Again, this relies on a bi-partition of the provers into two sets, such that, an encoded state shared across the bi-partition is equivalent to a Bell pair.

Let us first explain the general idea of the Pauli braiding test for self-testing n Bell pairs and n-qubit observables. We have a referee that is interacting with two players, labelled Alice and Bob. The test consists of three subtests which are chosen at random by the referee. The subtests are:

- **Linearity test.** In this test, the referee will randomly pick a basis setting, W, from the set $\{\mathsf{X}, \mathsf{Z}\}$. She then randomly chooses two strings $\mathbf{a_1}, \mathbf{a_2} \in \{0,1\}^n$ and sends them to Alice. With equal probability, the referee takes $\mathbf{b_1}$ to be either $\mathbf{a_1}$, $\mathbf{a_2}$ or $\mathbf{a_1} \oplus \mathbf{a_2}$. She also randomly chooses a string $\mathbf{b_2} \in \{0,1\}^n$ and sends the pair $(\mathbf{b_1}, \mathbf{b_2})$ to Bob[41]. Alice and Bob are then asked to measure the observables $W(\mathbf{a_1})$, $W(\mathbf{a_2})$ and $W(\mathbf{b_1})$, $W(\mathbf{b_2})$, respectively, on their

[41]Note that pair can be either $(\mathbf{b_1}, \mathbf{b_2})$ or $(\mathbf{b_2}, \mathbf{b_1})$, so that Bob does not know which string is the one related to Alice's inputs.

shared state. We denote Alice's outcomes as a_1, a_2 and Bob's outcomes as b_1, b_2. If $\mathbf{b_1} = \mathbf{a_1}$ (or $\mathbf{b_1} = \mathbf{a_2}$, respectively), the referee checks that $b_1 = a_1$ (or $b_1 = a_2$, respectively). If $\mathbf{b_1} = \mathbf{a_1} \oplus \mathbf{a_2}$, she checks that $b_1 = a_1 a_2$. This test is checking, on the one hand, that when Alice and Bob measure the same observables, they should get the same outcome (which is what should happen if they share Bell states). On the other hand, and more importantly, it is checking the commutation and linearity of their operators, i.e., that $W(\mathbf{a_1})W(\mathbf{a_2}) = W(\mathbf{a_2})W(\mathbf{a_1}) = W(\mathbf{a_1} + \mathbf{a_2})$ (and similarly for Bob's operators).

- **Anticommutation test.** The referee randomly chooses two strings $\mathbf{x}, \mathbf{z} \in \{0,1\}^n$, such that $\mathbf{x} \cdot \mathbf{z} = 1 \; mod \; 2$, and sends them to both players. These strings define the observables $\mathsf{X}(\mathbf{x})$ and $\mathsf{Z}(\mathbf{z})$ which are anticommuting because of the imposed condition on $\mathbf{x}$ and $\mathbf{z}$. The referee then engages in a non-local game with Alice and Bob designed to test the anticommutation of these observables for both of their systems. This can be any game that tests this property, such as the CHSH game or the *magic square* game, described in [78, 79]. As an example, if the referee chooses to play the CHSH game, then Alice will be instructed to measure either $\mathsf{X}(\mathbf{x})$ or $\mathsf{Z}(\mathbf{z})$ on her half of the shared state, while Bob would be instructed to measure either $(\mathsf{X}(\mathbf{x}) + \mathsf{Z}(\mathbf{z}))/\sqrt{2}$ or $(\mathsf{X}(\mathbf{x}) - \mathsf{Z}(\mathbf{z}))/\sqrt{2}$. The test is passed if the players achieve the win condition of the chosen anticommutation game. Note that for the case of the magic square game, the condition can be achieved with probability 1 when the players implement the optimal quantum strategy. For this reason, if the chosen game is the magic square game, then $\omega^*(PBT) = 1$.
- **Consistency test.** This test combines the previous two. The referee randomly chooses a basis setting, $W \in \{\mathsf{X}, \mathsf{Z}\}$ and two strings $\mathbf{x}, \mathbf{z} \in \{0,1\}^n$. Additionally, let $\mathbf{w} = \mathbf{x}$, if $W = \mathsf{X}$ and $\mathbf{w} = \mathbf{z}$ if $W = \mathsf{Z}$. The referee sends W, $\mathbf{x}$ and $\mathbf{z}$ to Alice. With equal probability the referee will then choose to perform one of two subtests. In the first subtest, the referee sends $\mathbf{x}, \mathbf{z}$ to Bob as well and plays the anticommutation game with both, such that Alice's observable is $W(\mathbf{w})$. As an example, if $W = \mathsf{X}$ and the game is the CHSH game, then Alice would be instructed to measure $\mathsf{X}(\mathbf{x})$, while Bob is instructed to measure either $(\mathsf{X}(\mathbf{x}) + \mathsf{Z}(\mathbf{z}))/\sqrt{2}$ or $(\mathsf{X}(\mathbf{x}) - \mathsf{Z}(\mathbf{z}))/\sqrt{2}$. This subtest essentially mimics the anticommutation test and is passed if the players achieve the win condition of the game. In the second subtest, which mimics the linearity test, the referee sends W, $\mathbf{w}$ and a random string $\mathbf{y} \in \{0,1\}^n$ to Bob, instructing him to measure $W(\mathbf{w})$ and $W(\mathbf{y})$. Alice is instructed to measure $W(\mathbf{x})$ and $W(\mathbf{z})$. The test if passed if Alice and Bob obtain the same result for the $W(\mathbf{w})$ observable. For instance, if $W = \mathsf{X}$, then both Alice and Bob will measure $\mathsf{X}(\mathbf{x})$ and their outcomes for that measurement must agree.

Having observables that satisfy the linearity conditions of the test as well as the anticommutation condition implies that they are isometric to the actual X and Z

Generator	Name
$II I\mathsf{XXXX}$	g_1
$I\mathsf{XX}II\mathsf{XX}$	g_2
$\mathsf{X}I\mathsf{X}I\mathsf{X}I\mathsf{X}$	g_3
$III\mathsf{ZZZZ}$	g_4
$I\mathsf{XX}II\mathsf{XZ}$	g_5
$\mathsf{Z}I\mathsf{Z}I\mathsf{Z}I\mathsf{Z}$	g_6

TABLE 5. Generators for Steane's 7-qubit code

observables acting on a maximally entangled state. This is what the Pauli braiding test checks and what is proven by the self-testing result of Natarajan and Vidick.

We can now describe the NV protocol. Similar to the FH protocol, for a quantum circuit, $\mathcal{C}$, and an input, x, one considers the associated Feynman-Kitaev state, denoted $|\psi\rangle$. This is then used to construct a 2-local XZ-Hamiltonian such that the ground state of this Hamiltonian is $|\psi\rangle$. As before, for some a and b, with $b - a > 1/poly(|\mathcal{C}|)$, when $\mathcal{C}$ accepts x we have that $\langle\psi| H |\psi\rangle < a$, otherwise $\langle\psi| H |\psi\rangle > b$. The verifier will instruct 7 provers to share a copy of $|\psi\rangle$ state, encoded in a 7-qubit quantum error correcting code known as *Steane's code*. The provers are then asked to perform measurements so as to self-test an encoded state or perform an energy measurement on this state. The code space, for Steane's code, is the 7-qubit subspace stabilized by all operators generated by $\{g_i\}_{i=1}^{6}$, where the generators are listed in Table 5.

The reason Natarajan and Vidick use this specific error correcting code is because it has two properties that are necessary for the application of their self-testing result. The first property is that each stabilizer generator is a tensor product of only the I, X and Z operators. This, of course, is true for the 5-qubit code as well. The second property is a symmetry condition: for each index $i \in \{1, \ldots, 6\}$, there exists a pair of stabilizer generators, S_{X} and S_{Z}, such that S_{X} consists exclusively of I and X operators and has an X on position i, whereas S_{Z} is identical to S_{X} but with all X operators replaced with Z. This property is not satisfied by the 5-qubit code and will allow the verifier to delegate to the provers measurements of the form $\mathsf{X}(\mathbf{x})$ and $\mathsf{Z}(\mathbf{z})$, where $\mathbf{x}$ and $\mathbf{z}$ are binary strings, as in the Pauli braiding test.

Putting everything together, the protocol works as follows. The verifier instructs the 7 provers to share an encoded instance of the Feynman-Kitaev state, $|\psi\rangle_L$, such that, for each logical qubit in $|\psi\rangle_L$, each prover will hold one of 7 the constituent physical qubits. She then chooses at random to perform one of the following:

- **Pauli braiding test.** The verifier chooses one of the 7 provers at random to be Alice, while the remaining provers will take on the role of Bob. The verifier then performs the Pauli braiding test with Alice and Bob in order to self-test the logical qubits in $|\psi\rangle_L$. As mentioned, each logical qubit, encoded in the 7 qubit code, is equivalent to a Bell pair under the chosen bi-partition. The Pauli braiding test is essentially checking that the provers have correctly encoded each of the qubits in $|\psi\rangle$ and that they are correctly measuring X and Z observables. The verifier rejects if the provers do not pass the test.

- **Energy test.** In this case, the verifier will pick an XZ-term, S, from H, at random, and instruct the provers to measure this term on their shared state. Note that S consists of *logical* X and Z operators. This means that each prover will need to perform local measurements so that their joint measurement acts as either X_L or Z_L, respectively. Additionally, X_L and Z_L, for the 7 qubit code, are expressed as tensor products of physical X and Z operations. This means that each prover will be instructed to measure an operators of the form $\mathsf{X}(\mathbf{x})$ and $\mathsf{Z}(\mathbf{z})$, on its physical qubits, where $\mathbf{x}, \mathbf{z} \in \{0,1\}^N$, and N is the total number of logical qubits on which H acts. The product $\mathsf{X}(\mathbf{x})\mathsf{Z}(\mathbf{z})$ is the outcome for that prover's share of S. The verifier then takes all of these ± 1 outcomes and multiplies them together, thus obtaining the outcome of measuring S itself. She accepts if the outcome of the measurement indicates that the estimated energy of $|\psi\rangle$ is below a and rejects otherwise.
- **Energy consistency test.** This test is a combination of the previous two. As in the Pauli braiding test, the provers are bi-partitioned into two sets, one consisting of one prover, denoted Alice, and the other consisting of the other 6 provers, jointly denoted as Bob. The verifier now performs a test akin to the linearity test from Pauli braiding. She randomly chooses $W \in \{\mathsf{X}, \mathsf{Z}\}$, and let $\mathbf{w} = \mathbf{x}$, if $W = \mathsf{X}$ and $\mathbf{w} = \mathbf{z}$ if $W = \mathsf{Z}$. She also chooses $\mathbf{x}, \mathbf{z} \in \{0,1\}^N$ according to the same distribution as in the energy test (i.e., as if she were instructing the provers to measure a random XZ-term from H). The verifier then does one of the following:
  - With probability $1/2$, instructs Alice to measure the observables $\mathsf{X}(\mathbf{x})$ and $\mathsf{Z}(\mathbf{z})$. Additionally, the verifier chooses $\mathbf{y} \in \{0,1\}^N$ at random and instructs Bob to measure $W(\mathbf{y})$ and $W(\mathbf{y} \oplus \mathbf{w})$. If $W = \mathsf{X}$, the verifier accepts if the product of Bob's answers agrees with Alice's answer for the $\mathsf{X}(\mathbf{x})$ observable. If $W = \mathsf{Z}$, the verifier accepts if the product of Bob's answers agrees with Alice's answer for the $\mathsf{Z}(\mathbf{z})$ observable. Note that this is the case since the product of Bob's observables should be $W(\mathbf{w})$ if he is behaving honestly.
  - With probability $1/4$, instructs Alice to measure $W(\mathbf{y})$ and $W(\mathbf{v})$, where $\mathbf{y}, \mathbf{w} \in \{0,1\}^N$ are chosen at random. Bob is instructed to measure $W(\mathbf{y})$ and $W(\mathbf{y} \oplus \mathbf{w})$. The verifier accepts if the outcomes of Alice and Bob for $W(\mathbf{y})$ agree.

– With probability 1/4, instructs Alice to measure $W(\mathbf{y} \oplus \mathbf{w})$ and $W(\mathbf{v})$, where $\mathbf{y}, \mathbf{w} \in \{0,1\}^N$ are chosen at random. Bob is instructed to measure $W(\mathbf{y})$ and $W(\mathbf{y} \oplus \mathbf{w})$. The verifier accepts if the outcomes of Alice and Bob for $W(\mathbf{y} \oplus \mathbf{w})$ agree.

The self-testing result guarantees that if these tests succeed, the verifier obtains an estimate for the energy of the ground state. Importantly, unlike the FH protocol, her estimate has constant precision. However, the protocol, as described up to this point, will still have an inverse polynomial completeness-soundness gap given by the local Hamiltonian. Recall that this is because the Feynman-Kitaev state will have energy below a when $\mathcal{C}$ accepts x with high probability, and energy above b otherwise, where $b - a > 1/|\mathcal{C}|^2$. But one can easily boost the protocol to a constant gap between completeness and soundness by simply requiring the provers to share $M = O(|\mathcal{C}|^2)$ copies of the ground state. This new state, $|\psi\rangle^{\otimes M}$, would then be the ground state of a new Hamiltonian H'[42]. One then runs the NV protocol for this Hamiltonian. It should be mentioned that this Hamiltonian is no longer 2-local, however, all of the tests in the NV protocol apply for these general Hamiltonians as well (as long as each term is comprised of I, X and Z operators, which is the case for H'). Additionally, the new Hamiltonian has a constant gap. The protocol therefore achieves a constant number of rounds of interaction with the provers (2 rounds) and we have that:

Theorem 16. *The NV protocol is an* MIP^* *protocol achieving a constant gap between completeness and soundness.*

To then boost the completeness-soundness gap to $1 - \epsilon$, for some $\epsilon > 0$, one can perform a parallel repetition of the protocol $O(\log(1/\epsilon))$ times.

4.4. Summary of Entanglement-based Protocols

We have seen that having non-communicating provers sharing entangled states allows for verification protocols with a classical client. What all of these protocols have in common is that they all make use of self-testing results. These essentially state that if a number of non-communicating players achieve a near optimal win rate in a non-local game, the strategy they employ in the game is essentially fixed, up to a local isometry. The strategy of the players consists of their shared quantum state as well as their local observables. Hence, self-testing results provide a precise characterisation for both.

This fact is exploited by the surveyed protocols in order to achieve verifiability. Specifically, we have seen that one approach is to define a number of non-local games so that by combining the optimal strategies of these games, the provers effectively perform a universal quantum computation. This is the approach employed by the RUV protocol [19]. Alternatively, the self-testing result can be used to check only for the correct preparation of a specific resource state. This resource state is then used by the provers to perform a quantum computation. How this is done

[42]Note that the state still needs to be encoded in the 7 qubit code.

depends on the type of resource state and on how the computation is delegated to the provers. For instance, one possibility is to remotely prepare the resource state used in the VUBQC protocol and then run the verification procedure of that protocol. This is the approach used by the GKW and HPDF protocols [20, 21]. Another possibility is to prepare a cluster state shared among many provers and then have each of those provers measure their states so as to perform an MBQC computation. This approach was used by McKague in his protocol [22]. Lastly, the self-tested resource state can be the ground state of a local Hamiltonian leading to the post-hoc approaches employed by the FH and NV protocols.

Protocol	Provers	Qmem provers	Rounds	Communication	Blind
RUV	2	2	$O(N^{8192} \cdot \log(1/\epsilon))$	$O(N^{8192} \cdot \log(1/\epsilon))$	Y
McKague	$O(N^{22} \cdot \log(1/\epsilon))$	0	$O(N^{22} \cdot \log(1/\epsilon))$	$O(N^{22} \cdot \log(1/\epsilon))$	Y
GKW	2	1	$O(N^{2048} \cdot \log(1/\epsilon))$	$O(N^{2048} \cdot \log(1/\epsilon))$	Y
HPDF	$O(N^4 \log(N) \cdot \log(1/\epsilon))$	$O(\log(1/\epsilon))$	$O(N^4 \log(N) \cdot \log(1/\epsilon))$	$O(N^4 \log(N) \cdot \log(1/\epsilon))$	Y
FH	5	5	$O(N^{16} \cdot \log(1/\epsilon))$	$O(N^{19} \cdot \log(1/\epsilon))$	N
NV	7	7	$O(1)$	$O(N^3 \cdot \log(1/\epsilon))$	N

TABLE 6. Comparison of entanglement-based protocols. We denote $N = |\mathcal{C}|$ to be the size of the delegated quantum computation together with the input to that computation. The listed values are given assuming a completeness-soundness gap of at least $1 - \epsilon$, for some $\epsilon > 0$. For the "Qmem provers" column, the numbers indicate how many provers need to have a quantum memory that is not of constant size, with respect to $|\mathcal{C}|$ (if we ignore the preparation of the initial shared entangled state). The "Rounds" column quantifies how many rounds of interaction are performed between the verifier and the provers, whereas "Communication" quantifies the total amount of communication (number of rounds times the size of the messages). Note that a similar table can be found in [25].

We noticed that, depending on the approach that is used, there will be different requirements for the quantum operations of the provers. Of course, all protocols require that collectively the provers can perform BQP computations, however, individually some provers need not be universal quantum computers. Related to this is the issue of blindness. Again, based on what approach is used some protocols utilize blindness and some do not. In particular, the post-hoc protocols are not blind since the computation and the input are revealed to the provers so that they can prepare the Feynman-Kitaev state.

We have also seen that the robustness of the self-testing game impacts the communication complexity of the protocol. Specifically, having robustness which is inverse polynomial in the number of qubits of the self-tested state, leads to an inverse polynomial gap between completeness and soundness. In order to make this gap constant, the communication complexity of the protocol has to be made polynomial. This means that most protocols will have a relatively large overhead, when compared to prepare-and-send or receive-and-measure protocols. Out of the surveyed protocols, the NV protocol is the only one which utilizes a self-testing result with constant robustness and therefore has a constant completeness-soundness gap. We summarize all of these facts in Table 6[43].

5. Outlook

5.1. Sub-universal Protocols

So far we have presented protocols for the verification of universal quantum computations, i.e., protocols in which the provers are assumed to be BQP machines. In the near future, however, quantum computers might be more limited in terms of the type of computations that they can perform. Examples of this include the class of so-called *instantaneous quantum computations*, denoted IQP, *boson sampling* or the *one-pure qubit model* of quantum computation [1, 2, 80]. While not universal, these examples are still highly relevant since, assuming some plausible complexity theoretic conjectures hold, they could solve certain problems or sample from certain distributions that are intractable for classical computers. One is therefore faced with the question of how to verify the correctness of outcomes resulting from these models. In particular, when considering an interactive protocol, the prover should be restricted to the corresponding sub-universal class of problems and yet still be able to prove statements to a computationally limited verifier. We will see that many of the considered approaches are adapted versions of the VUBQC protocol from Subsection 2.2. It should be noted, however, that the protocols themselves are not direct applications of VUBQC. In each instance, the protocol was constructed so as to adhere to the constraints of the model.

The first sub-universal verification protocol is for the one-pure (or one-clean) qubit model. A machine of this type takes as input a state of limited purity (for instance, a system comprising of the totally mixed state and a small number of single qubit pure states), and is able to coherently apply quantum gates. The model was considered in order to reflect the state of a quantum computer with noisy storage. In [81], Kapourniotis, Kashefi and Datta introduced a verification protocol for this model by adapting VUBQC to the one-pure qubit setting. The verifier still prepares individual pure qubits, as in the original VUBQC protocol, however the prover holds a mixed state of limited purity at all times[44]. Additionally, the prover can inject or remove pure qubits from his state, during the computation, as long as it does not increase the total purity of the state. The resulting protocol has

---

[43]Note that for the HPDF protocol we assumed that there is one prover with quantum memory, comprised of the individual provers that come together in order to perform the MBQC computation at the end of the protocol. Since, to achieve a completeness-soundness gap of $1 - \epsilon$, the protocol is repeated $O(\log(1/\epsilon))$ times, this means there will be $O(\log(1/\epsilon))$ provers with quantum memory in total.

[44]The purity of a d-qubit state, ρ, is quantified by the *purity parameter* defined in [81] as: $\pi(\rho) = \log(Tr(\rho^2)) + d$.

an inverse polynomial completeness-soundness gap. However, unlike the universal protocols we have reviewed, the constraints on the prover's state do not allow for the protocol to be repeated. This means that the completeness-soundness gap cannot be boosted through repetition.

Another model, for which verification protocols have been proposed, is that of instantaneous quantum computations, or IQP [2, 82]. An IQP machine is one which can only perform unitary operations that are diagonal in the X basis and therefore commute with each other. The name "instantaneous quantum computation" illustrates that there is no temporal structure to the quantum dynamics [2]. Additionally, the machine is restricted to measurements in the computational basis. It is important to mention that IQP does not represent a decision class, like BQP, but rather a *sampling class.* The input to a sampling problem is a specification of a certain probability distribution and the output is a sample from that distribution. The class IQP, therefore, contains all distributions which can be sampled efficiently (in polynomial time) by a machine operating as described above. Under plausible complexity theoretic assumptions, it was shown that this class is not contained in the set of distributions which can be efficiently sampled by a classical computer [82].

In [2], Shepherd and Bremner proposed a hypothesis test in which a classical verifier is able to check that the prover is sampling from an IQP distribution. The verifier cannot, however, check that the prover sampled from the correct distributions. Nevertheless, the protocol serves as a practical tool for demonstrating a quantum computational advantage. The test itself involves an encoding, or obfuscation scheme which relies on a computational assumption (i.e., it assumes that a particular problem is intractable for IQP machines).

Another test of IQP problems is provided by the Hangleiter et al approach, from Subsection 3.2 [31]. Recall that this was essentially the 1S-Post-hoc protocol for certifying the ground state of a local Hamiltonian. Hangleiter et al have the prover prepare multiple copies of a state which is the Feynman-Kitaev state of an IQP circuit. They then use the post-hoc protocol to certify that the prover prepared the correct state (measuring local terms from the Hamiltonian associated with that state) and then use one copy to sample from the output of the IQP circuit. This is akin to the measurement-only approach of Subsection 3.1. In a subsequent paper, by Bermejo-Vega et al, they consider a subclass of sampling problems that are contained in IQP and prove that this class is also hard to classically simulate (subject to standard complexity theory assumptions). The problems can be viewed as preparing a certain entangled state and then measuring all qubits in a fixed basis. The authors provide a way to certify that the state prepared is close to the ideal one, by giving an upper bound on the trace distance. Moreover, the measurements required for this state certification can be made using local stabilizer measurements, for the considered architectures and settings [5].

Recently, another scheme has been proposed, by Mills et al [83], which again adapts the VUBQC protocol to the IQP setting. This eliminates the need for computational assumptions, however it also requires the verifier to have a single qubit preparation device. In contrast to VUBQC, however, the verifier need only prepare eigenstates of the Y and Z operators.

Yet another scheme derived from VUBQC was introduced in [84] for a model known as the *Ising spin sampler*. This is based on the *Ising model*, which describes a lattice of interacting spins in the presence of a magnetic field [85]. The Ising spin sampler is a translation invariant Ising model in which one measures the spins thus obtaining samples from the partition function of the model. Just like with IQP, it was shown in [86] that, based on complexity theoretic assumptions, sampling from the partition function is intractable for classical computers.

Lastly, Disilvestro and Markham proposed a verification protocol [87] for *Spekkens' toy model* [88]. This is a local hidden variable theory which is phenomenologically very similar to quantum mechanics, though it cannot produce non-local correlations. The existence of the protocol, again inspired by VUBQC, suggests that Bell non-locality is not a necessary feature for verification protocols, at least in the setting in which the verifier has a trusted quantum device.

5.2. Fault Tolerance

The protocols reviewed in this paper have all been described in an ideal setting in which all quantum devices work perfectly and any deviation from the ideal behaviour is the result of malicious provers. This is not, however, the case in the real world. The primary obstacle, in the development of scalable quantum computers, is noise which affects quantum operations and quantum storage devices. As a solution to this problem, a number of fault tolerant techniques, utilizing quantum error detection and correction, have been proposed. Their purpose is to reduce the likelihood of the quantum computation being corrupted by imperfect gate operations. But while these techniques have proven successful in minimizing errors in quantum computations, it is not trivial to achieve the same effect for verification protocols. To clarify, while we have seen the use of quantum error correcting codes in verification protocols, their purpose was to either boost the completeness-soundness gap (in the case of prepare-and-send protocols), or to ensure an honest behaviour from the provers (in the case of entanglement-based post-hoc protocols). The question we ask, therefore, is: how can one design a fault-tolerant verification protocol? Note that this question pertains primarily to protocols in which the verifier is not entirely classical (such as the prepare-and-send or receive-and-measure approaches) or in which one or more provers are assumed to be single-qubit devices (such as the GKW and HPDF protocols). For the remaining entanglement-based protocols, one can simply assume that the provers are performing all of their operations on top of a quantum error correcting code.

Let us consider what happens if, in the prepare-and-send and receive-and-measure protocols, the devices of the verifier and the prover are subject to noise[45]. If, for simplicity, we assume that the errors on these devices imply that each qubit

[45]Different noise models have been examined when designing fault tolerant protocols, however, a very common model and one which can be considered in our case, is *depolarizing noise* [89,90]. This can be single-qubit depolarizing noise, which acts as $\mathcal{E}(\rho) = (1-p)[I] + p/3([\mathsf{X}] + [\mathsf{Y}] + [\mathsf{Z}])$,

will have a probability, p, of producing the same outcome as in the ideal setting, when measured, we immediately notice that the probability of n qubits producing the same outcomes scales as $O(p^n)$. This means that, even if the prover behaves honestly, the computation is very unlikely to result in the correct outcome [20].

Ideally, one would like the prover to perform his operations in a fault tolerant manner. In other words, the prover's state should be encoded in a quantum error correcting code, the gates he performs should result in logical operations being applied on his state and he should, additionally, perform error-detection (syndrome) measurements and corrections. But we can see that this is problematic to achieve. Firstly, in prepare-and-send protocols, the computation state of the prover is provided by the verifier. Who should then encode this state in the error-correcting code, the verifier or the prover? It is known that in order to suppress errors in a quantum circuit, $\mathcal{C}$, each qubit should be encoded in a logical state having $O(polylog(|\mathcal{C}|)$-many qubits [89]. This means that if the encoding is performed by the verifier, she must have a quantum computer whose size scales poly-logarithmically with the size of the circuit that she would like to delegate. It is preferable, however, that the verifier has a constant-size quantum computer. Conversely, even if the prover performs the encoding, there is another complication. Since the verifier needs to encrypt the states she sends to the prover, and since her operations are susceptible to noise, the errors acting on these states will have a dependency on her secret parameters. This means that when the prover performs error-detection procedures he could learn information about these secret parameters and compromise the protocol.

For receive-and-measure protocols, one encounters a different obstacle. While the verifier's measurement device is not actively malicious, if the errors occurring in this device are correlated with the prover's operations in preparing the state, this can compromise the correctness of the protocol.

A number of fault tolerant verification protocols have been proposed, however, they all overcome these limitations by making additional assumptions. For instance, one proposal, by Kapourniotis and Datta [84], for making VUBQC fault tolerant, uses a topological error-correcting code described in [57, 58]. The error-correcting code is specifically designed for performing fault tolerant MBQC computations, which is why it is suitable for the VUBQC protocol. In the proposed scheme, the verifier still prepares single qubit states, however there is an implicit assumption that the errors on these states are independent of the verifier's secret parameters. The prover is then instructed to perform a blind MBQC computation in the topological code. The protocol described in [84] is used for a specific type of MBQC computation designed to demonstrate a quantum computational advantage. However, the authors argue that the techniques are general and could be applied for universal quantum computations.

---

or two-qubit depolarizing noise, which acts as $\mathcal{E}(\rho) = (1-p)[I \otimes I] + p/15([I \otimes \mathsf{X}] + \ldots + [\mathsf{Z} \otimes \mathsf{Z}])$, for some probability $p > 0$. The square bracket notation indicates the action of an operator.

A fault-tolerant version of the measurement-only protocol from Subsection 3.1 has also been proposed in [91]. The graph state prepared by the prover is encoded in an error-correcting code, such as the topological lattice used by the previous approaches. As in the 'non-fault-tolerant' version of the protocol, the prover is instructed to send many copies of this state which the verifier will test using stabilizer measurements. The verifier also uses one copy in order to perform her computation in an MBQC fashion. The protocol assumes that the errors occurring on the verifier's measurement device are independent of the errors occurring on the prover's devices.

More details, regarding the difficulties with achieving fault tolerance in QPIP protocols, can be found in [27].

5.3. Experiments and Implementations

Protocols for verification will clearly be useful for benchmarking experiments implementing quantum computations. Experiments implementing quantum computations on a small number of qubits can be verified with brute force simulation on a classical computer. However, as we have pointed out that this is not scalable, in the long-term it is worthwhile to try and implement verification protocols on these devices. As a result, there have been proof of concept experiments that demonstrate the components necessary for verifiable quantum computing.

Inspired by the prepare-and-send VUBQC protocol, Barz et al implemented a four-photon linear optical experiment, where the four-qubit linear cluster state was constructed from entangled pairs of photons produced through parametric down-conversion [92]. Within this cluster state, in runs of the experiment, a trap qubit was placed in one of two possible locations, thus demonstrating some of the elements of the VUBQC protocol. However, it should be noted that the trap qubits are placed in the system through measurements on non-trap qubits within the cluster state, i.e., through measurements made on the the other three qubits. Because of this, the analysis of the VUBQC protocol cannot be directly translated over to this setting, and bespoke analysis of possible deviations is required. In addition, the presence of entanglement between the photons was demonstrated through Bell tests that are performed blindly. This work also builds on a previous experimental implementation of blind quantum computation by Barz et al [93].

With regards to receive-and-measure protocols, and in particular the measurement-only protocol of Subsection 3.1, Greganti et al implemented [94] some of the elements of these protocols with a four-photon experiment, similar to the experiment of Barz et al mentioned above [92]. This demonstration builds on previous work in the experimental characterisation of stabiliser states [95]. In this case, two four-qubit cluster states were generated: the linear cluster state and the star graph state, where in the latter case the only entanglement is between one central qubit and pairwise with every other qubit. In order to demonstrate the elements for measurement-only verification, by suitable measurements made by the client, traps can be placed in the state. Furthermore, the linear cluster state and star graph state can be used as computational resources for implementing single qubit unitaries and an entangling gate respectively.

Finally, preliminary steps have been taken towards an experimental implementation of the RUV protocol, from Subsection 4.1.1. Huang et al implemented a simplified version of this protocol using sources of pairs of entangled photons [71]. Repeated CHSH tests were performed on thousands of pairs of photons demonstrating a large violation of the CHSH inequality; a vital ingredient in the protocol of RUV. In between the many rounds of CHSH tests, state tomography, process tomography, and a computation were performed, with the latter being the factorisation of the number 15. Again, all of these elements are ingredients in the protocol, however, the entangled photons are created "on-the-fly". In other words, in RUV, two non-communicating provers share a large number of maximally entangled states prior to the full protocol, but in this experiment these states are generated throughout.

6. Conclusions

The realization of the first quantum computers capable of outperforming classical computers at non-trivial tasks is fast approaching. All signs indicate that their development will follow a similar trajectory to that of classical computers. In other words, the first generation of quantum computers will comprise of large servers that are maintained and operated by specialists working either in academia, industry or a combination of both. However, unlike with the first super-computers, the Internet opens up the possibility for users, all around the world, to interface with these devices and delegate problems to them. This has already been the case with the 5-qubit IBM machine [96], and more powerful machines are soon to follow [97,98]. But how will these computationally restricted users be able to verify the results produced by the quantum servers? That is what the field of quantum verification aims to answer. Moreover, as mentioned before and as is outlined in [12], the field also aims to answer the more foundational question of: how do we verify the predictions of quantum mechanics in the large complexity regime?

In this paper, we have reviewed a number of protocols that address these questions. While none of them achieve the ultimate goal of the field, which is to have a classical client verify the computation performed by a single quantum server, each protocol provides a unique approach for performing verification and has its own advantages and disadvantages. We have seen that these protocols combine elements from a multitude of areas including: cryptography, complexity theory, error correction and the theory of quantum correlations. We have also seen that proof-of-concept experiments, for some of these protocols, have already been realized.

What all of the surveyed approaches have in common, is that none of them are based on computational assumptions. In other words, they all perform verification unconditionally. However, recently, there have been attempts to reduce the verifier's requirements by incorporating computational assumptions as well. What this means is that the protocols operate under the assumption that certain problems are intractable for quantum computers. We have already mentioned an example: a protocol for verifying the sub-universal sampling class of IQP computations, in which the verifier is entirely classical. Other examples include protocols for *quantum fully homomorphic encryption* [99, 100]. In these protocols, a client is delegating a quantum computation to a server while trying to keep the input to the computation hidden. The use of computational assumptions allows these protocols to achieve this functionality using only one round of back-and-forth communication. However, in the referenced schemes, the client does require some minimal quantum capabilities. A recent modification of these schemes has been proposed in order to make the protocols verifiable as well [101]. Additionally, an even more recent paper introduces a protocol for quantum fully homomorphic encryption with an entirely classical client (again, based on computational assumptions) [102]. We can therefore see a new direction emerging in the field of delegated quantum computations. This recent success in developing protocols based on computational assumptions could very well lead to the first single-prover verification protocol with a classical client.

Another new direction, especially pertaining to entanglement-based protocols, is given by the development of self-testing results achieving constant robustness. This started with the work of Natarajan and Vidick, which was the basis of their protocol from Subsection 4.3.2 [24]. We saw, in Section 4, that all entanglement-based protocols rely, one way or another, on self-testing results. Consequently, the robustness of these results greatly impacts the communication complexity and overhead of these protocols. Since most protocols were based on results having inverse polynomial robustness, this led to prohibitively large requirements in terms of quantum resources (see Table 6). However, subsequent work by Coladangelo et al, following up on the Natarajan and Vidick result, has led to two entanglement-based protocols, which achieve near linear overhead [25][46]. This is a direct consequence of using a self-testing result with constant robustness and combining it with the Test-or-Compute protocol of Broadbent from Subsection 2.3. Of course, of the two protocols proposed by Coladangelo et al, only one is blind and so an open problem, of their result, is whether the second protocol can also be made blind. Another question is whether the protocols can be further optimized so that only one prover is required to perform universal quantum computations, in the spirit of the GKW protocol from Subsection 4.1.2.

We conclude by listing a number of other open problems that have been raised by the field of quantum verification. The resolution of these problems is relevant not just to quantum verification but to quantum information theory as a whole.

---

[46]The result from [25] appeared on the arxiv close to the completion of this work, which is why we did not review it.

- While the problem of a classical verifier delegating computations to a single prover is the main open problem of the field, we emphasize a more particular instance of this problem: can the proof that any problem in PSPACE[47] admits an interactive proof system, be adapted to show that any problem in BQP admits an interactive proof system with a BQP prover? The proof that PSPACE = IP (in particular the PSPACE $\subseteq$ IP direction) uses error-correcting properties of low-degree polynomials to give a verification protocol for a PSPACE-complete problem [103]. We have seen that the Poly-QAS VQC scheme, presented in Subsection 2.1.2, also makes use of error-correcting properties of low-degree polynomials in order to perform quantum verification (albeit, with a quantum error correcting code and a quantum verifier). Can these ideas lead to a classical verifier protocol for BQP problems with a BQP prover?
- In all existing entanglement-based protocols, one assumes that the provers are not allowed to communicate during the protocol. However, this assumption is not enforced by physical constraints. Is it, therefore, possible to have an entanglement-based verification protocol in which the provers are *space-like separated*[48]? Note, that since all existing protocols require the verifier to query the two (or more) provers adaptively, it is not directly possible to make the provers be space-like separated.
- What is the optimal overhead (in terms of either communication complexity, or the resources of the verifier) in verification protocols? For all types of verification protocols we have seen that, for a fixed completeness-soundness gap, the best achieved communication complexity is linear. For the prepare-and-send case is it possible to have a protocol in which the verifier need only prepare a poly-logarithmic number of single qubits (in the size of the computation)? For the entanglement-based case, can the classical verifier send only poly-logarithmic sized questions to the provers? This latter question is related to the quantum PCP conjecture [104].
- Are there other models of quantum computation that are suitable for developing verification protocols? We have seen that the way in which we view quantum computations has a large impact on how we design verification protocols and what characteristics those protocols will have. Specifically, the separation between classical control and quantum resources in MBQC lead to VUBQC, or the QMA-completeness of the local Hamiltonian problem lead to the post-hoc approaches. Of course, all universal models are equivalent in

[47]PSPACE is the class of problems which can be solved in polynomial space by a classical computer.

[48]In an experiment, two regions are space-like separated if the time it takes light to travel from one region to the other is longer than the duration of the experiment. Essentially, according to relativity, this means that there is no causal ordering between events occurring in one region and events occurring in the other.

terms of the computations which can be performed, however each model provides a particular insight into quantum computation which can prove useful when devising new protocols. Can other models of quantum computation, such as the adiabatic model, the anyon model etc, provide new insights?

- We have seen that while certain verification protocols employ error-correcting codes, these are primarily used for boosting the completeness-soundness gap. Alternatively, for the protocols that do in fact incorporate fault tolerance, in order to cope with noisy operations, there are additional assumptions such as the noise in the verifier's device being uncorrelated with the noise in the prover's devices. Therefore, the question is: can one have a fault tolerant verification protocol, with a minimal quantum verifier, in the most general setting possible? By this we mean that there are no restrictions on the noise affecting the quantum devices in the protocol, other than those resulting from the standard assumptions of fault tolerant quantum computations (constant noise rate, local errors etc). This question is addressed in more detail in [27]. Note that the question refers in particular to prepare-and-send and receive-and-measure protocols, since entanglement-based approaches are implicitly fault tolerant (one can assume that the provers are performing the computations on top of error correcting codes).

Acknowledgements

The authors would like to thank Petros Wallden, Alex Cojocaru, Thomas Vidick for very useful comments and suggestions for improving this work, and Dan Mills for TEX support. AG would also like to especially thank Matty Hoban for many helpful remarks and comments and Vivian Uhlir for useful advice in improving the figures in the paper. EK acknowledges funding through EPSRC grant EP/N003829/1 and EP/M013243/1. TK acknowledges funding through EPSRC grant EP/K04057X/2.

7. Appendix

7.1. Quantum Information and Computation

In this section, we provide a few notions regarding the basics of quantum information and quantum computation and refer the reader to the appropriate references for a more in depth presentation [89, 105, 106].

7.1.1. Basics of quantum mechanics. A quantum state (or a quantum register) is a unit vector in a complex Hilbert space, $\mathcal{H}$. We denote quantum states, using standard Dirac notation, as $|\psi\rangle \in \mathcal{H}$, called a 'ket' state. The dual of this state is denoted $\langle\psi|$, called a 'bra', and is a member of the dual space $\mathcal{H}^{\perp}$. We will only be concerned with finite-dimensional Hilbert spaces. Qubits are states in two-dimensional Hilbert spaces. Traditionally, one fixes an orthonormal basis for such a space, called *computational basis*, and denotes the basis vectors as $|0\rangle$ and $|1\rangle$. Gluing together systems to express the states of multiple qubits is achieved through *tensor product*, denoted $\otimes$. The notation $|\psi\rangle^{\otimes n}$ denotes a state comprising of n copies of $|\psi\rangle$. If a state $|\psi\rangle \in \mathcal{H}_1 \otimes \mathcal{H}_2$ cannot be expressed as $|a\rangle \otimes |b\rangle$, for any $|a\rangle \in \mathcal{H}_1$ and any $|b\rangle \in \mathcal{H}_2$, we say that the state is *entangled*. As a shorthand, we will sometimes write $|a\rangle\,|b\rangle$ instead of $|a\rangle \otimes |b\rangle$. As a simple example of an entangled state one can consider the *Bell state*:

$$|\Phi_+\rangle = \frac{|00\rangle + |11\rangle}{\sqrt{2}}\,. \tag{69}$$

Quantum mechanics postulates that there are two ways to change a quantum state: *unitary evolution* and *measurement*. Unitary evolution involves acting with some unitary operation U on $|\psi\rangle$, thus producing the mapping $|\psi\rangle \to U\,|\psi\rangle$. Note that any such operation is reversible through the application of the hermitian conjugate of U, denoted $U^\dagger$, since $UU^\dagger = U^\dagger U = I$.

Measurement, in its most basic form, involves expressing a state $|\psi\rangle$ in a particular orthonormal basis, $\mathcal{B}$, and then choosing one of the basis vectors as the state of the system post-measurement. The index of that vector is the classical outcome of the measurement. The post-measurement vector is chosen at random and the probability of obtaining a vector $|v\rangle \in \mathcal{B}$ is given by $|\,\langle v|\psi\rangle\,|^2$.

More generally, a measurement involves a collection of operators $\{M_i\}_i$ acting on the state space of the system to be measured and satisfying:

1. $M_i M_i^\dagger = M_i^\dagger M_i\,$;
2. $M_i M_i^\dagger$ is a positive operator;
3. $\sum_i M_i M_i^\dagger = I\,$.

The label i indicates a potential measurement outcome. Given a state $|\psi\rangle$ to be measured, the probability of obtaining outcome i is $p(i) = \langle\psi|\, M_i^\dagger M_i\,|\psi\rangle$ and the state of the system after the measurement will be $M_i\,|\psi\rangle\,/\sqrt{p(i)}$. If we are only interested in the probabilities of the different outcomes and not in the post-measurement state then we can denote $E_i = M_i M_i^\dagger$ and we will refer to the set $\{E_i\}_i$ as a *positive-operator valued measure* (POVM). When performing a measurement in a basis $\mathcal{B} = \{|i\rangle\}_i$, we are essentially choosing $M_i = |i\rangle\,\langle i|$. This is known as a *projective measurement* and in general consists of operators M_i satisfying the property that $M_i^2 = M_i$.

Lastly, when discussing measurements we will sometimes use *observables*. These are hermitian operators which define a measurement specified by the diagonal basis of the operator. Specifically, for some hermitian operator O, we know that there exists a basis $\mathcal{B} = \{|i\rangle\}_i$ such that:

$$O = \sum_i \lambda_i\,|i\rangle\,, \tag{70}$$

where $\{\lambda_i\}_i$ is the set of eigenvalues of O. Measuring the O observable on some state $|\psi\rangle$ is equivalent to performing a projective measurement of $|\psi\rangle$ in the basis

$\mathcal{B}$[49]. When using observables, one takes the measurement outcomes to be the eigenvalues of O, rather than the basis labels. In other words, if when measuring O the state is projected to $|i\rangle$, then the measurement outcome is taken to be λ_i.

7.1.2. Density matrices. States denoted by kets are also referred to as *pure states*. Quantum mechanics tells us that for an isolated quantum system the complete description of that system is given by a pure state[50]. This is akin to classical physics where pure states are points in phase space, which provide a complete characterisation of a classical system. However, unlike classical physics where knowing the pure state uniquely determines the outcomes of all possible measurements of the system, in quantum mechanics measurements are probabilistic even given the pure state. It is also possible that the state of a quantum system is specified by a probability distribution over pure states. This is known as a *mixed state* and can be represented using *density matrices*. These are positive semidefinite, trace one, hermitian operators.

The density matrix of a pure state $|\psi\rangle$ is $\rho = |\psi\rangle \langle\psi|$. For an ensemble of states $\{|\psi_i\rangle\}_i$, each occurring with probability p_i, such that $\sum_i p_i = 1$, the corresponding density matrix is:

$$\rho = \sum_i p_i |\psi_i\rangle \langle\psi_i| \,. \tag{71}$$

It can be shown that if ρ corresponds to a pure state then $Tr(\rho^2) = 1$, whereas when ρ is a mixed state $Tr(\rho^2) < 1$. One of the most important mixed states, which we encounter throughout this review, is the *maximally mixed state*. The density matrix for this state is I/d, where I is the identity matrix and d is the dimension of the underlying Hilbert space. As an example, the maximally mixed state for a one qubit system is $I/2$. This state represents the state of maximal uncertainty about quantum system. What this means is that for any basis $\{|v_i\rangle\}_i$ of the Hilbert space of dimension d, the maximally mixed state is:

$$\frac{I}{d} = \frac{1}{d} \sum_{i=1}^{d} |v_i\rangle \langle v_i| \,. \tag{72}$$

Equivalently, any projective measurement, specified by a complete basis $\mathcal{B}$, of the maximally mixed state will have all outcomes occurring with equal probability. We will denote the set of all density matrices over some Hilbert space $\mathcal{H}$ as $\mathcal{D}(\mathcal{H})$.

When performing a measurement on a state ρ, specified by operators $\{M_i\}_i$, the probability of outcome i is given by $p(i) = Tr(M_i M_i^\dagger \rho)$ and the post-measurement state will be $M_i \rho M_i^\dagger / p(i)$.

---

[49]Note that if the operator is degenerate (i.e., has repeating eigenvalues) then the projectors for degenerate eigenvalues will correspond to projectors on the subspaces spanned by the associated eigenvectors.

[50]It should be noted that this is the case provided that quantum mechanics is a *complete* theory in terms of its characterisation of physical systems. See [107] for more details.

7.1.3. Purification. An essential operation concerning density matrices is the *partial trace*. This provides a way of obtaining the density matrix of a subsystem that is part of a larger system. Partial trace is linear, and is defined as follows. Given two density matrices ρ_1 and ρ_2 with Hilbert spaces $\mathcal{H}_1$ and $\mathcal{H}_2$, we have that:

$$\rho_1 = Tr_2(\rho_1 \otimes \rho_2), \qquad \rho_2 = Tr_1(\rho_1 \otimes \rho_2). \tag{73}$$

In the first case one is 'tracing out' system 2, whereas in the second case we trace out system 1. This property together with linearity completely defines the partial trace. For if we take any general density matrix, ρ, on $\mathcal{H}_1 \otimes \mathcal{H}_2$, expressed as:

$$\rho = \sum_{i,i',j,j'} a_{ii'jj'} \left|i\right\rangle_1 \left\langle i'\right|_1 \otimes \left|j\right\rangle_2 \left\langle j'\right|_2, \tag{74}$$

where $\{|i\rangle\}$ ($\{|i'\rangle\}$) and $\{|j\rangle\}$ ($\{|j'\rangle\}$) are orthonormal bases for $\mathcal{H}_1$ and $\mathcal{H}_2$, if we would like to trace out subsystem 2, for example, we would then have:

$$Tr_2(\rho) = Tr_2\left(\sum_{i,i',j,j'} a_{ii'jj'} \left|i\right\rangle_1 \left\langle i'\right|_1 \otimes \left|j\right\rangle_2 \left\langle j'\right|_2 \right) = \sum_{i,i',j} a_{ii'jj} \left|i\right\rangle_1 \left\langle i'\right|_1. \tag{75}$$

An important fact, concerning the relationship between mixed states and pure states, is that any mixed state can be *purified*. In other words, for any mixed state ρ over some Hilbert space $\mathcal{H}_1$ one can always find a pure state $|\psi\rangle \in \mathcal{H}_1 \otimes \mathcal{H}_2$ such that $dim(\mathcal{H}_1) = dim(\mathcal{H}_2)$[51] and:

$$Tr_2(|\psi\rangle\langle\psi|) = \rho. \tag{76}$$

Moreover, the purification $|\psi\rangle$ is not unique and so another important result is the fact that if $|\phi\rangle \in \mathcal{H}_1 \otimes \mathcal{H}_2$ is another purification of ρ then there exists a unitary U, acting only on $\mathcal{H}_2$ (the additional system that was added to purify ρ) such that:

$$|\phi\rangle = (I \otimes U)\,|\psi\rangle. \tag{77}$$

We will refer to this as the *purification principle.*

7.1.4. CPTP maps. All operations on quantum states can be viewed as maps from density matrices on an input Hilbert space to density matrices on an output Hilbert space, $\mathcal{O} : \mathcal{D}(\mathcal{H}_{in}) \to \mathcal{D}(\mathcal{H}_{out})$, which may or may not be of the same dimension. Quantum mechanics dictates that such a map, must satisfy three properties:

1. **Linearity**: $\mathcal{O}(a\rho_1 + b\rho_2) = a\mathcal{O}(\rho_1) + b\mathcal{O}(\rho_2)$.
2. **Complete positivity**: the map $\mathcal{O} \otimes I : \mathcal{D}(\mathcal{H}_{in} \otimes \mathcal{H}_E) \to \mathcal{D}(\mathcal{H}_{out} \otimes \mathcal{H}_E))$ takes positive states to positive states, for all extensions $\mathcal{H}_E$.
3. **Trace preserving**: $Tr(\mathcal{O}(\rho)) = Tr(\rho)$.

For this reason, such maps are referred to as *completely positive trace-preserving* (CPTP) maps. It can be shown that any CPTP map can be equivalently expressed as:

$$\mathcal{O}(\rho) = \sum_i K_i \rho K_i^\dagger \tag{78}$$

[51] One could allow for purifications in larger systems, but we restrict attention to same dimensions.

for some set of linear operators $\{K_i\}_i$, known as *Kraus operators*, satisfying:

$$\sum_i K_i K_i^\dagger = I \tag{79}$$

CPTP maps are also referred to as *quantum channels*. Additionally, we also mention *isometries* which are CPTP maps $\mathcal{O}$ for which $\mathcal{O}^\dagger \circ \mathcal{O} = I$.

7.1.5. Trace distance. We will frequently be interested in comparing the "closeness" of quantum states. To do so we will use the notion of *trace distance* which generalizes *variation distance* for probability distributions. Recall that if one has two probability distributions $p(x)$ and $q(x)$, over a finite sample space, the variation distance between them is defined as:

$$D(p, q) = \frac{1}{2} \sum_x |p(x) - q(x)| . \tag{80}$$

Informally, this represents the largest possible difference between the probabilities that the two distributions can assign to some even x. The quantum analogue of this, for density matrices, is:

$$TD(\rho_1, \rho_2) = \frac{1}{2} Tr \left(\sqrt{(\rho_1 - \rho_2)(\rho_1 - \rho_2)^\dagger} \right) . \tag{81}$$

One could think that the trace distance simply represents the variation distance between the probability distributions associated with measuring ρ_1 and ρ_2 in the same basis (or using the same POVM). However, there are infinitely many choices of a measurement basis. So, in fact, the trace distance is the *maximum* over all possible measurements of the variation distance between the corresponding probability distributions.

Similar to variation distance, the trace distance takes values between 0 and 1, with 0 corresponding to identical states and 1 to perfectly distinguishable states. Additionally, like any other distance measure, it satisfies the triangle inequality.

7.1.6. Quantum computation. Quantum computation is most easily expressed in the *quantum gates model*. In this framework, gates are unitary operations which act on groups of qubits. As with classical computation, universal quantum computation is achieved by considering a fixed set of quantum gates which can approximate any unitary operation up to a chosen precision. The most common universal set of gates is given by:

$$\mathsf{X} = \begin{bmatrix} 0 & 1 \\ 1 & 0 \end{bmatrix}, \mathsf{Z} = \begin{bmatrix} 1 & 0 \\ 0 & -1 \end{bmatrix}, \mathsf{H} = \frac{1}{\sqrt{2}} \begin{bmatrix} 1 & 1 \\ 1 & -1 \end{bmatrix}, \mathsf{T} = \begin{bmatrix} 1 & 0 \\ 0 & e^{i\pi/4} \end{bmatrix},$$
$$\mathsf{CNOT} = \begin{bmatrix} 1 & 0 & 0 & 0 \\ 0 & 1 & 0 & 0 \\ 0 & 0 & 0 & 1 \\ 0 & 0 & 1 & 0 \end{bmatrix}. \tag{82}$$

In order, the operations are known as Pauli X and Pauli Z, Hadamard, the T-gate and controlled-NOT. Note that general controlled-U operations are operations performing the mapping $|0\rangle|\psi\rangle \to |0\rangle|\psi\rangle$, $|1\rangle|\psi\rangle \to |1\rangle U|\psi\rangle$. The first qubit is known as a *control qubit*, whereas the second is known as *target qubit*. The matrices express the action of each operator on the computational basis. A classical outcome for a particular quantum computation can be obtained by measuring the quantum state resulting from the application of a sequence of quantum gates. Another gate, which we will encounter, is the *Toffoli* gate, or the controlled-controlled-NOT gate, described by the matrix:

$$\mathsf{CCNOT} = \begin{bmatrix} 1 & 0 & 0 & 0 & 0 & 0 & 0 & 0 \\ 0 & 1 & 0 & 0 & 0 & 0 & 0 & 0 \\ 0 & 0 & 1 & 0 & 0 & 0 & 0 & 0 \\ 0 & 0 & 0 & 1 & 0 & 0 & 0 & 0 \\ 0 & 0 & 0 & 0 & 1 & 0 & 0 & 0 \\ 0 & 0 & 0 & 0 & 0 & 1 & 0 & 0 \\ 0 & 0 & 0 & 0 & 0 & 0 & 0 & 1 \\ 0 & 0 & 0 & 0 & 0 & 0 & 1 & 0 \end{bmatrix}. \quad (83)$$

The effect of this gate is to apply an X on a target qubit if both control qubits are in the $|1\rangle$ state.

We also mention an important class of quantum operations known as *Clifford operations*. To define them, consider first the n-qubit Pauli group:

$$\mathbb{P}_n = \{\alpha\, \sigma_1 \otimes \ldots \otimes \sigma_n | \alpha \in \{+1, -1, +i, -i\}, \sigma_i \in \{I, \mathsf{X}, \mathsf{Y}, \mathsf{Z}\}\}. \quad (84)$$

As a useful side note, the n-qubit Pauli group forms a basis for all $2^n \times 2^n$ matrices. The *Clifford group* is then defined as follows:

$$\mathfrak{C}_n = \{U \in U(2^n) | \sigma \in P_n \implies U\sigma U^\dagger \in P_n\}. \quad (85)$$

Where $U(2^n)$ is the set of all $2^n \times 2^n$ unitary matrices. Clifford operations, therefore, are operations which leave the Pauli group invariant under conjugation. Operationally they can be obtained through combinations of the Pauli gates together with H, CNOT and $\mathsf{S} = \mathsf{T}^2$, in which case they are referred to as *Clifford circuits*. We note that the T and Toffoli gates are not Clifford operations. However, Clifford circuits combined with either of these two gates gives a universal set of quantum operations.

7.1.7. Bloch sphere. The final aspect we mention is the *Bloch sphere*, which offers a useful geometric picture for visualizing single qubit states. Any such state is represented as a point on the surface of the sphere. In Figure 16, one can see a visualization of the Bloch sphere together with the states $|0\rangle, |1\rangle$, the eigenstates of Z, as well as $|+\rangle = \frac{1}{\sqrt{2}}(|0\rangle + |1\rangle), |-\rangle = \frac{1}{\sqrt{2}}(|0\rangle - |1\rangle)$, the eigenstates of X and $|+_{\pi/2}\rangle = \frac{1}{\sqrt{2}}(|0\rangle + i|1\rangle), |-_{\pi/2}\rangle = \frac{1}{\sqrt{2}}(|0\rangle - i|1\rangle)$, the eigenstates of Y. All of the previously mentioned single-qubit operations can be viewed as rotations on this sphere. The Pauli $\mathsf{X}, \mathsf{Y}, \mathsf{Z}$ gates correspond to rotations by π radians around the corresponding $\mathsf{X}, \mathsf{Y}, \mathsf{Z}$ axes. The Hadamard gate, which can be expressed as $\mathsf{H} = \frac{1}{\sqrt{2}}(\mathsf{X} + \mathsf{Z})$ acts as a rotation by π radians around the $\mathsf{X} + \mathsf{Z}$ axis. Lastly, the T gate, corresponds to a rotation by $\pi/4$ radians around the Z axis.

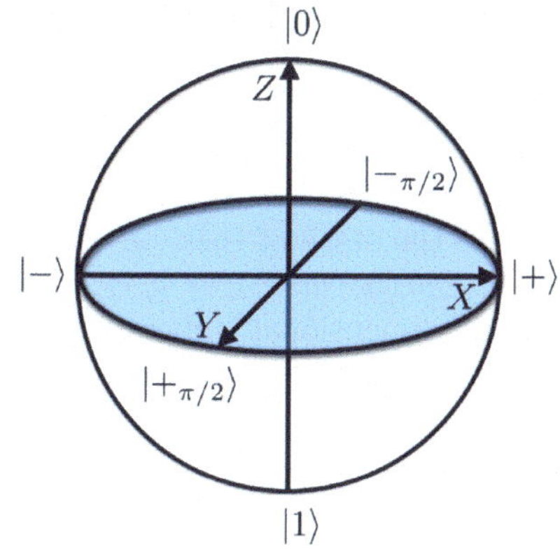

FIGURE 16. Bloch sphere

We will frequently mention the states $|+_\phi\rangle = \frac{1}{\sqrt{2}}(|0\rangle + e^{i\phi}|1\rangle)$ and $|-_\phi\rangle = \frac{1}{\sqrt{2}}(|0\rangle - e^{i\phi}|1\rangle)$ which all lie in the XY-plane of the Bloch sphere, represented in blue in the above figure. These states can be viewed as rotations of the $|+\rangle, |-\rangle$ states by ϕ radians around the Z axis. For example, the $|+_{\pi/2}\rangle, |-_{\pi/2}\rangle$ states are rotations by $\pi/2$ around the Z axis of the $|+\rangle, |-\rangle$ states. One can also consider measurements in the XY-plane. Any two diametrically opposed states in this plane form a basis for a one-qubit Hilbert space and therefore define a projective measurement. Suppose we choose the basis $(|+_\phi\rangle, |-_\phi\rangle)$ and wish to measure the state $|+_\theta\rangle$. It can be shown that the probability of the state being projected to $|+_\phi\rangle$ is $cos^2((\phi - \theta)/2)$, whereas the probability of it being projected to $|-_\phi\rangle$ is $sin^2((\phi - \theta)/2)$. In other words, the probabilities only depend on the *angle difference* between ϕ and θ. This fact will prove very useful later on.

7.1.8. Quantum error correction. One important consideration, when discussing quantum protocols, is that any implementation of quantum operations will be subject to noise stemming from interactions with the external environment. For this reason, one needs a *fault tolerant* way of performing quantum computation. This is achieved using protocols for quantum error detection and correction, for which we give a simplified description.

Suppose we have a k-qubit quantum state $|\psi\rangle$ on which we want to perform some quantum gate G. The quantum memory storing $|\psi\rangle$ as well as the implementation of G are subject to noise. This means that if we were to apply G directly on $|\psi\rangle$ the result would be $\mathcal{E}(G|\psi\rangle)$, where $\mathcal{E}$ is a CPTP error map associated with the noisy application of G. Using the Kraus decomposition, the action of $\mathcal{E}$ can be expressed as:

$$\mathcal{E}(G|\psi\rangle) = \sum_j E_j \, G|\psi\rangle\langle\psi| G^\dagger \, E_j^\dagger, \tag{86}$$

where $\{E_j\}_j$ is a set of Kraus operators. If one can correct for all E_j's then one can correct for $\mathcal{E}$ as well [108].

To detect and correct for errors from the set $\{E_j\}_j$, one first performs an encoding procedure on $|\psi\rangle$ mapping it to a so-called *logical state* $|\psi\rangle_L$ on n qubits, where $n > k$. This procedure involves the use of $n - k$ auxiliary qubits known as *ancilla* qubits. If we denote the state of these $n - k$ ancillas as $|anc\rangle$ we then have the encoding procedure $Enc(|\psi\rangle |anc\rangle) \rightarrow |\psi\rangle_L$. This state is part of a 2^k-dimensional subspace of the 2^n-dimensional Hilbert space of all n qubits, denoted $\mathcal{H}$. The subspace is usually referred to as the *code space* of the error correcting code. One way to represent this space is by giving a set of operators such that the code space is the intersection of the $+1$ eigenspaces of all the operators.

As an example, consider the 3-qubit *flip code*. We will take $k = 1$ and $n = 3$, so that one qubit is encoded in 3 qubits. The code is able to detect and correct for Pauli X errors occurring on *a single* qubit. The encoding procedure for a state $|\psi\rangle = a|0\rangle + b|1\rangle$ maps it to the state $|\psi\rangle_L = a|000\rangle + b|111\rangle$. The code space is therefore defined by $span(|000\rangle, |111\rangle)$. It is also the unique $+1$ eigenspace of the operators $g_1 = Z \otimes Z \otimes I$ and $g_2 = I \otimes Z \otimes Z$[52]. All valid operations on $|\psi\rangle_L$ must be invariant on this subspace, whereas any error from the set $\{E_j\}_j$ should map the state to a different subspace. In this case, valid operations, or *logical operations*, are the analogues of the single-qubit unitaries that map $|\psi\rangle \rightarrow |\phi\rangle = U|\psi\rangle$. Thus, a logical operation U_L would map $|\psi\rangle_L \rightarrow |\phi\rangle_L$. The error set simply consists of $\{\mathsf{X} \otimes I \otimes I, I \otimes \mathsf{X} \otimes I, I \otimes I \otimes \mathsf{X}\}$. We can see that any of these errors will map a state inside $span(|000\rangle, |111\rangle)$ to a state outside of this code space. One then defines a projective measurement in which the projectors are associated with each of the 2^{n-k} subspaces of $\mathcal{H}$. This is called a *syndrome measurement*. Its purpose is to detect whether an error has occurred and, if so, which error. Knowing this, the effect of the error can be undone by simply applying the inverse operation. For the 3-qubit code, there are $2^{3-1} = 4$ possible subspaces in which the state can be mapped to, meaning that we need a 4-outcome measurement. The syndrome measurement is defined by jointly measuring the observables g_1 and g_2. An outcome of $+1$ for both observables indicates that the state is in the correct subspace, $span(|000\rangle, |111\rangle)$. Conversely, if either of the two observables produces a -1 outcome, then this corresponds to one of the 3 possible errors. For instance, an outcome of $+1$ for the first observable and -1 for the second, indicates that the state is in the subspace $span(|001\rangle, |110\rangle)$, corresponding to an X error on the third qubit. The error is corrected by applying another X operation on that qubit.

Since Kraus operators can be expressed in terms of Paulis acting on the individual qubits, one often speaks about the *weight* of an error correcting code. If the code can correct non-identity Pauli operations on at most w qubits, then w is the weight of the code.

---

[52]These are known as *stabilizer* operators for the states in the code spaces. We also encounter these operators in Subsection 7.2. The operators form a group under multiplication and so, when specifying the code space, it is sufficient to provide the generators of the group.

The smallest error correcting code which can correct for *any* single-qubit error is the 5-*qubit code* (i.e., one qubit is encoded as 5 qubits) [76]. This code is used Subsection 4.3.

7.2. Measurement-based Quantum Computation

Since some of the protocols we review are expressed in the model of *measurement-based quantum computation* (MBQC), defined in [109, 110], we provide a brief description of this model.

Unlike the quantum gates model of computation, in MBQC a given computation is performed by measuring qubits from a large entangled state. Traditionally, this state consists of qubits prepared in the state $|+\rangle = \frac{1}{\sqrt{2}}(|0\rangle + |1\rangle)$, entangled using the CZ (controlled-Z) operation, where:

$$\mathsf{CZ} = \begin{bmatrix} 1 & 0 & 0 & 0 \\ 0 & 1 & 0 & 0 \\ 0 & 0 & 1 & 0 \\ 0 & 0 & 0 & -1 \end{bmatrix}.$$

They are then measured in the basis $(|+_\phi\rangle, |-_\phi\rangle)$. These measurements are denoted as $M(\phi)$, and depending on the value of ϕ chosen for each qubit one can perform universal quantum computation. For this to work, the entangled qubits need to form a *universal graph state*. Such a state consists, for example, of N qubits in the state $|+\rangle$. These qubits have been entangled according to some graph structure G, such that there exist measurement patterns (an order for measuring the qubits in G and the corresponding measurement angles) for each quantum computation consisting of $O(N)$ gates. We will denote the graph state of the $|+\rangle$ qubits entangled according to the structure of G as $|G\rangle$.

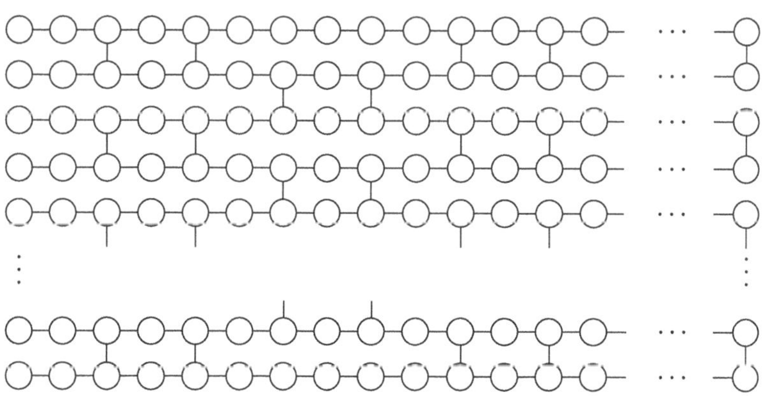

FIGURE 17. Brickwork state, reproduced from [37]

A universal graph state allows one to perform any quantum computation up to a certain size. An example of such a state is the *brickwork state*, defined in [37] from which we illustrate Figure 17. To be more precise, suppose we would like to perform some quantum computation described by a circuit consisting of N gates. The corresponding MBQC computation consists of the following steps:

1. **Initialization**. Prepare $O(N)$ qubits, each in the state $|+\rangle$.
2. **Entanglement**. Entangle the qubits according to some universal graph state structure, such as the brickwork state.
3. **Measurement**. Measure each qubit, i using $M(\phi_i)$, for some angle ϕ_i determined based on the computation we would like to perform. The angles ϕ_i are referred to as the *computation angles*.
4. **Correction**. Apply appropriate corrections (Pauli X and Z operations) to the qubits, based on the measurement outcomes.

The last two steps can be performed together. This is because if we would like to apply a Pauli X correction to a qubit, i, before measuring it, we can simply measure it using $M(-\phi_i)$. Similarly, if we would like to apply a Pauli Z correction to that same qubit we measure it using $M(\phi_i + \pi)$. Therefore, the general measurement performed on a particular qubit will be $M((-1)^s \phi_i + r\pi)$, where $s, r \in \{0, 1\}$ are determined by previous measurement outcomes.

One element concerning graph states, which we will encounter in some protocols, is the representation of these states using *stabilizers*. A stabilizer state for a unitary hermitian operator, O, is some state $|\psi\rangle$ such that $O|\psi\rangle = |\psi\rangle$. O is referred to as a stabilizer of $|\psi\rangle$. It is possible to specify a state, $|\psi\rangle$, by giving a set of operators, such that $|\psi\rangle$ is the unique state which is stabilized by all the operators in the set. As an example, the state $|\Phi_+\rangle = (|00\rangle + |11\rangle)/\sqrt{2}$ is uniquely stabilized by the set $\{\mathsf{X} \otimes \mathsf{X}, \mathsf{Z} \otimes \mathsf{Z}\}$. Note that the set of all stabilizers for a state forms a group, since if $O_1|\psi\rangle = |\psi\rangle$ and $O_2|\psi\rangle = |\psi\rangle$, then clearly $O_1 O_2 |\psi\rangle = |\psi\rangle$. So, it is sufficient to specify a set of generators for that group in order to describe the stabilizer group for a particular state.

To specify the generators for the stabilizer group of a graph state $|G\rangle$, let us first denote $V(G)$ as the set of vertices in the graph G and $N_G(v)$ as the set of neighbouring vertices for some vertex v (i.e., all vertices in G that are connected to v through an edge). Additionally, for some operator O, when we write O_v we mean that O is acting on the qubit from $|G\rangle$ associated with vertex v in G. The generators for the stabilizer group of $|G\rangle$ are then given by:

$$K_v = \mathsf{X}_v \prod_{w \in N_G(v)} \mathsf{Z}_w \tag{87}$$

for all $v \in V(G)$.

As a final remark, it should be noted that one can translate quantum circuits into MBQC patterns in a canonical way. For instance, the universal gate set mentioned in the previous subsection, and hence any quantum circuit comprising of those gates, can be translated directly into MBQC. See for example [37] for more details.

7.3. Complexity Theory

As mentioned in the introduction, the questions regarding verification of quantum computation can be easily expressed in the language of complexity theory. To that end, we provide definitions for the basic complexity classes used in this paper. We let $\{0,1\}^*$ denote the set of all binary strings and $\{0,1\}^n$ the set of all binary strings of length n. We use standard complexity theory notation and assume familiarity with the concepts of Turing machines and uniform circuits. For a more general introduction into the subject we refer the reader to [111, 112].

Definition 5. *A language $L \subseteq \{0,1\}^*$ belongs to* BPP *if there exist polynomials p, and a probabilistic Turing machine M, whose running time on inputs of size n is bounded by $p(n)$, such that for any $x \in \{0,1\}^n$ the following is true:*

- *when $x \in L$, $M(x)$*[53] *accepts with probability at least c,*
- *when $x \notin L$, $M(x)$ accepts with probability at most s,*

where $c - s \geq 1/p(n)$.

Here, and in all subsequent definitions, c is referred to as *completeness* and s is referred to as *soundness*. Traditionally, one takes $c = 2/3$ and $s = 1/3$, however, in full generality, the only requirement is that there exists an inverse polynomial gap between c and s.

Definition 6. *A language $L \subseteq \{0,1\}^*$ belongs to* BQP *if there exist polynomials p, and a uniform quantum circuit family $\{C_n\}_n$, where each circuit has at most $p(n)$ gates, such that for any $x \in \{0,1\}^n$ the following is true:*

- *when $x \in L$, $C_n(x)$ accepts with probability at least c,*
- *when $x \notin L$, $C_n(x)$ accepts with probability at most s,*

where $c - s \geq 1/p(n)$.

For the quantum circuit C_n, acceptance can be defined as having one of its output qubits outputting 1 when measured in the computational basis.

Definition 7. *A language $L \subseteq \{0,1\}^*$ belongs to* MA *if there exist polynomials p, and a probabilistic Turing machine V, whose running time on inputs of size n is bounded by $p(n)$, such that for any $x \in \{0,1\}^n$ the following is true:*

- *when $x \in L$, there exists a string $w \in \{0,1\}^{\leq p(n)}$, such that $V(x,w)$ accepts with probability at least c,*
- *when $x \notin L$, for all strings $w \in \{0,1\}^{\leq p(n)}$, $V(x,w)$ accepts with probability at most s,*

where $c - s \geq 1/p(n)$.

For this class, V is traditionally referred to as the verifier (or Arthur), whereas w, which is the witness string, is provided by the prover (or Merlin). Essentially, the verifier and is tasked with checking a purported proof that $x \in L$, provided by the prover. There is also a quantum version of this class:

[53] The notation $M(x)$ means running the Turing machine M on input x.

Definition 8. *A language $L \subseteq \{0,1\}^*$ belongs to* QMA *if there exists a polynomial p and a uniform quantum circuit family $\{V_n\}_n$ taking x and a quantum state $|\psi\rangle$ as inputs, such that for any $x \in \{0,1\}^n$ the following are true:*

- *when $x \in L$, there exists a quantum state $|\psi\rangle \in \mathcal{H}$, such that $V_n(x, |\psi\rangle)$ accepts with probability at least c, and*
- *when $x \notin L$, for all quantum states $|\psi\rangle \in \mathcal{H}$, $V_n(x, |\psi\rangle)$ accepts with probability at most s,*

where $dim(\mathcal{H}) \leq 2^{p(|x|)}$ and $c - s \geq 1/p(|x|)$.

For QMA we also provide the definition of a complete problem[54] since this will be referenced in some of the protocols we review. The specific problem we state was defined by Kitaev et al and is known as the *k-local Hamiltonian problem* [63]. A k-local Hamiltonian, acting on a system of n qubits, is a hermitian operator H that can be expressed as $H = \sum_i H_i$, where each H_i is a hermitian operator which acts non-trivially on at most k qubits. We give the definition of the k-local Hamiltonian problem from [104]:

Definition 9 (The k-local Hamiltonian (LH) problem).

- ***Input:*** *$H_1, \ldots, H_m$, a set of m Hermitian matrices each acting on k qubits out of an n-qubit system and satisfying $\|H_i\| \leq 1$. Each matrix entry is specified by $poly(n)$-many bits. Apart from the H_i we are also given two real numbers, a and b (again, with polynomially many bits of precision) such that $\Gamma = b - a > 1/poly(n)$. Γ is referred to as the* absolute promise gap *of the problem.*
- ***Output:*** *Is the smallest eigenvalue of $H = H_1 + H_2 + \ldots + H_m$ smaller than a or are all its eigenvalues larger than b?*

Essentially, for some language $L \in$ QMA, and given a and b, one can construct a k-local Hamiltonian such that, whenever $x \in L$, its smallest eigenvalue is less than a and whenever $x \notin L$, all of its eigenvalues are greater than b. The witness $|\psi\rangle$, when $x \in L$, is the eigenstate of H corresponding to its lowest eigenvalue (or one such eigenstate if the Hamiltonian is degenerate). The uniform circuit family $\{V_n\}_n$ represents a BQP verifier, whereas the state $|\psi\rangle$ is provided by a prover. The verifier receives this witness from the prover and measures one of the local terms H_i (which is an observable) on that state. This can be done with a polynomial-size quantum circuit and yields an estimate for measuring H itself. Therefore, when $x \in L$ and the prover sends $|\psi\rangle$, with high probability the verifier will obtain the corresponding eigenvalue of $|\psi\rangle$ which will be smaller than a. Conversely, when $x \notin L$, no matter what state the prover sends, with high probability, the verifier will measure a value above b. The constant k, in the definition, is not arbitrary. In the initial construction of Kitaev, k had to be at least 5 for the problem to be QMA-complete. Subsequent work has shown that even with $k = 2$ the problem remains QMA-complete [64].

[54] A problem, P, is complete for the complexity class QMA if $P \in$ QMA and all problems in QMA can be reduced in quantum polynomial time to P.

Definition 10. *A language $L \subseteq \{0,1\}^*$ belongs to* IP *if there exist polynomials p, and a probabilistic Turing machine V, whose running time on inputs of size n is bounded by $p(n)$, such that for any $x \in \{0,1\}^n$ the following is true:*

- *when $x \in L$, there exists a prover P which exchanges at most $p(n)$ messages (of length at most $p(n)$) with V and makes V accept with probability at least c,*
- *when $x \notin L$, any prover P which exchanges at most $p(n)$ messages (of length at most $p(n)$) with V, makes V accept with probability at most s,*

where $c - s \geq 1/p(n)$.

While the previous are fairly standard complexity classes, we now state the definition of a more non-standard class, which first appeared in [26]:

Definition 11. *A language $L \subseteq \{0,1\}^*$ belongs to* QPIP *if there exist polynomials p, a constant κ and a probabilistic Turing machine V, whose running time on inputs of size n is bounded by $p(n)$, and which is augmented with the ability to prepare and measure groups of κ qubits, such that for any $x \in \{0,1\}^n$ the following is true:*

- *when $x \in L$, there exists a* BQP *prover P which exchanges at most $p(n)$ classical or quantum messages (of length at most $p(n)$) with V and makes V accept with probability at least c,*
- *when $x \notin L$, any* BQP *prover P which exchanges at most $p(n)$ classical or quantum messages (of length at most $p(n)$) with V, makes V accept with probability at most s,*

where $c - s \geq 1/p(n)$.

Some clarifications are in order. The class QPIP differs from IP in two ways. Firstly, while computationally the verifier is still restricted to the class BPP, operationally it has the additional ability of preparing or measuring groups of κ qubits. Importantly, κ is a constant which is independent of the size of the input. This is why this extra ability does not add to the verifier's computational power, since a constant-size quantum device can be simulated in constant time by a BPP machine. Secondly, unlike IP, in QPIP the prover is restricted to BQP computations. This constraint on the prover is more in line with Problem 1 and it also has the direct implication that QPIP $\subseteq$ BQP.

As we will see, all the protocols in Section 2 and Section 3 are QPIP protocols. And since these protocols allow for the delegation of arbitrary BQP problems, it follows that QPIP $=$ BQP.

We now proceed to the multi-prover setting and define the multi-prover generalization of IP:

Definition 12. *A language $L \subseteq \{0,1\}^*$ belongs to* MIP[k] *if there exist polynomials p, and a probabilistic Turing machine V, whose running time on inputs of size n is bounded by $p(n)$, such that for any $x \in \{0,1\}^n$ the following is true:*

- *when* $x \in L$, *there exists a* k-*tuple of provers* $(P_1, P_2, \ldots, P_k)$ *which are not allowed to communicate and which exchange at most* $p(n)$ *messages (of length at most* $p(n)$*) with* V *and make* V *accept with probability at least* c,
- *when* $x \notin L$, *any* k-*tuple of provers* $(P_1, P_2, \ldots, P_k)$ *which are not allowed to communicate and which exchange at most* $p(n)$ *messages (of length at most* $p(n)$*) with* V, *make* V *accept with probability at most* s,

where $c - s \geq 1/p(n)$.

Note that MIP[1] = IP and it was shown that for all $k > 2$, MIP[k] = MIP[2] [113]. The latter class is simply denoted MIP. If the provers are allowed to share entanglement then we obtain the class:

Definition 13. *A language* $L \subseteq \{0,1\}^*$ *belongs to* MIP*[k] *if there exist polynomials* p, *and a probabilistic Turing machine* V, *whose running time on inputs of size* n *is bounded by* $p(n)$, *such that for any* $x \in \{0,1\}^n$ *the following is true:*

- *when* $x \in L$, *there exists a* k-*tuple of provers* $(P_1, P_2, \ldots, P_k)$ *which can share arbitrarily many entangled qubits, are not allowed to communicate and which exchange at most* $p(n)$ *messages (of length at most* $p(n)$*) with* V *and make* V *accept with probability at least* c,
- *when* $x \notin L$, *any* k-*tuple of provers* $(P_1, P_2, \ldots, P_k)$ *which can share arbitrarily many entangled qubits, are not allowed to communicate and which exchange at most* $p(n)$ *messages (of length at most* $p(n)$*) with* V, *make* V *accept with probability at most* s,

where $c - s \geq 1/p(n)$.

As before it is the case that MIP*[k] = MIP*[2] and this class is denoted as MIP* [114]. It is not known whether MIP = MIP*, however, it is known that both classes contain BQP. Importantly, for MIP* protocols, if the provers are restricted to BQP computations, the resulting complexity class is equal to BQP [19]. Most of the protocols presented in Section 4 are of this type.

Note that while the protocols we review can be understood in terms of the listed complexity classes, we will often give a more fine-grained description of their functionality and resources than is provided by complexity theory. To give an example, for a QPIP protocol, from the complexity theoretic perspective, we are interested in the verifier's ability to delegate arbitrary BQP decision problems to the prover by interacting with it for a polynomial number of rounds. In practice, however, we are interested in a number of other characteristics of the protocol such as:

- whether the verifier can delegate not just decision problems, but also sampling problems (i.e., problems in which the verifier wishes to obtain a sample from a particular probability distribution and is able to certify that, with high probability, the sample came from the correct distribution),
- whether the prover can receive a particular quantum input for the computation or return a quantum output to the verifier,
- having minimal quantum communication between the verifier and the prover,

– whether the verifier can "hide" the input and output of the computation from the prover.

References

[1] S. Aaronson and A. Arkhipov, The computational complexity of linear optics. In: *Proceedings of the Forty-third Annual ACM Symposium on Theory of Computing*, STOC '11, New York, NY, USA, ACM (2011), 333–342.

[2] D. Shepherd and M.J. Bremner, Temporally unstructured quantum computation. In: *Proceedings of the Royal Society of London A: Mathematical, Physical and Engineering Sciences* **465** (2009), 1413–1439.

[3] S. Boixo, S.V. Isakov, V.N. Smelyanskiy, R. Babbush, N. Ding, Z. Jiang, J.M. Martinis, and H. Neven, Characterizing quantum supremacy in near-term devices. arXiv:1608.00263 (2016).

[4] S. Aaronson and L. Chen, Complexity-theoretic foundations of quantum supremacy experiments. arXiv:1612.05903 (2016).

[5] J. Bermejo-Vega, D. Hangleiter, M. Schwarz, R. Raussendorf, and J. Eisert, Architectures for quantum simulation showing a quantum speedup (2017).

[6] M. Tillmann, B. Dakić, R. Heilmann, S. Nolte, A. Szameit, and P. Walther, Experimental boson sampling. *Nature Photonics* **7** (7) (2013), 540–544.

[7] N. Spagnolo, C. Vitelli, M. Bentivegna, D.J. Brod, A. Crespi, F. Flamini, S. Giacomini, G. Milani, R. Ramponi, P. Mataloni, et al., Experimental validation of photonic boson sampling. *Nature Photonics* **8** (8) (2014), 615–620.

[8] M. Bentivegna, N. Spagnolo, C. Vitelli, F. Flamini, N. Viggianiello, L. Latmiral, P. Mataloni, D.J. Brod, E.F. Galvão, A. Crespi, et al., Experimental scattershot boson sampling. *Science advances* **1** (3) (2015), e1400255.

[9] B. Lanyon, M. Barbieri, M. Almeida, and A. White, Experimental quantum computing without entanglement. *Physical review letters* **101** (20) (2008), 200501.

[10] S. Aaronson, The Aaronson $25.00 prize. `http://www.scottaaronson.com/blog/?p=284`.

[11] U. Vazirani, Workshop on the computational worldview and the sciences. `http://users.cms.caltech.edu/~schulman/Workshops/CS-Lens-2/report-comp-worldview.pdf` (2007).

[12] D. Aharonov and U. Vazirani, Is quantum mechanics falsifiable? A computational perspective on the foundations of quantum mechanics. In: *Computability: Turing, Gödel, Church, and Beyond*. MIT Press (2013).

[13] R. Impagliazzo and A. Wigderson, P= bpp if e requires exponential circuits: Derandomizing the xor lemma. In: *Proceedings of the twenty-ninth annual ACM symposium on Theory of computing*, ACM (1997), 220–229.

[14] P.W. Shor, Polynomial-time algorithms for prime factorization and discrete logarithms on a quantum computer. *SIAM Review* **41** (2) (1999), 303–332

[15] D. Aharonov and I. Arad, The bqp-hardness of approximating the jones polynomial. *New Journal of Physics* **13** (3) (2011), 035019.

[16] T.O. Yeates, T.S. Norcross, and N.P. King, Knotted and topologically complex proteins as models for studying folding and stability. *Current opinion in chemical biology* **11** (6) (2007), 595–603.

[17] E. Witten, Quantum field theory and the jones polynomial. *Communications in Mathematical Physics* **121** (3) (1989), 351–399.

[18] A.M. Childs, R. Cleve, E. Deotto, E., Farhi, S. Gutmann, and D.A. Spielman, Exponential algorithmic speedup by a quantum walk. In: *Proceedings of the thirty-fifth annual ACM symposium on Theory of computing*, ACM (2003), 59–68.

[19] B.W. Reichardt, F. Unger, and U. Vazirani, Classical command of quantum systems. *Nature* **496** (7446) (2013), 456.

[20] A. Gheorghiu, E. Kashefi, and P. Wallden, Robustness and device independence of verifiable blind quantum computing. *New Journal of Physics* **17** (8) (2015), 083040.

[21] M. Hajdušek, C.A. Pérez-Delgado, and J.F. Fitzsimons, Device-independent verifiable blind quantum computation. arXiv:1502.02563 (2015).

[22] M. McKague, Interactive proofs for BQP via self-tested graph states. *Theory of Computing* **12** (3) (2016), 1–42.

[23] J.F. Fitzsimons and M. Hajdušek, Post hoc verification of quantum computation. arXiv:1512.04375 (2015).

[24] A. Natarajan, and T. Vidick, Robust self-testing of many-qubit states. arXiv:1610.03574 (2016).

[25] A. Coladangelo, A. Grilo, S. Jeffery, and T. Vidick, Verifier-on-a-leash: new schemes for verifiable delegated quantum computation, with quasilinear resources. arXiv:1708.07359 (2017).

[26] D. Aharonov, M. Ben-Or, and E. Eban, Interactive proofs for quantum computations. In: *Innovations in Computer Science* – ICS 2010, Tsinghua University, Beijing, China, January 5–7, 2010. Proceedings (2010), 453–469.

[27] D. Aharonov, M. Ben-Or, E. Eban, and U. Mahadev, Interactive proofs for quantum computations. arXiv:1704.04487 (2017).

[28] J.F. Fitzsimons and E. Kashefi, Unconditionally verifiable blind quantum computation. *Phys. Rev. A* **96** (Jul 2017), 012303.

[29] A. Broadbent, How to verify a quantum computation. arXiv:1509.09180 (2015).

[30] T. Morimae and J.F. Fitzsimons, Post hoc verification with a single prover. arXiv:1603.06046 (2016).

[31] D. angleiter, M. Kliesch, M. Schwarz, and J. Eisert, Direct certification of a class of quantum simulations. *Quantum Science and Technology* **2** (1) (2017), 015004.

[32] M. Hayashi and T. Morimae, Verifiable measurement-only blind quantum computing with stabilizer testing. *Physical review letters* **115** (22) (2015), 220502.

[33] T. Morimae, Y. Takeuchi, and M. Hayashi, Verified measurement-based quantum computing with hypergraph states. arXiv:1701.05688 (2017).

[34] A. Gheorghiu, P. Wallden, and E. Kashefi, Rigidity of quantum steering and one-sided device-independent verifiable quantum computation. *New Journal of Physics* **19** (2) (2017), 023043.

[35] J.F. Fitzsimons, Private quantum computation: an introduction to blind quantum computing and related protocols. *npj Quantum Information* **3** (1) (2017), 23.

[36] A.M. Childs, Secure assisted quantum computation. *Quantum Info. Comput.* **5** (6) (September 2005), 456–466.

[37] A. Broadbent, J. Fitzsimons, and E. Kashefi, Universal blind quantum computation. In: *Proceedings of the 50th Annual Symposium on Foundations of Computer Science.* FOCS '09, IEEE Computer Society (2009), 517–526.

[38] P. Arrighi and L. Salvail, Blind quantum computation. *International Journal of Quantum Information* **04** (05) (2006), 883–898.

[39] V. Giovannetti, L. Maccone, T. Morimae, and T.G. Rudolph, Efficient universal blind quantum computation. *Phys. Rev. Lett.* **111** (Dec 2013), 230501.

[40] A. Mantri, C.A. Pérez-Delgado, and J.F. Fitzsimons, Optimal blind quantum computation. *Phys. Rev. Lett.* **111** (Dec 2013), 230502.

[41] R.L. Rivest, L. Adleman, and M.L. Dertouzos, On data banks and privacy homomorphisms. *Foundations of secure computation* **4** (11) (1978), 169–180.

[42] C. Gentry, Fully homomorphic encryption using ideal lattices. In: *Proceedings of the Forty-first Annual ACM Symposium on Theory of Computing.* STOC '09, New York, NY, USA, ACM (2009), 169–178.

[43] Z. Brakerski and V. Vaikuntanathan, Efficient fully homomorphic encryption from (standard) LWE. In: *Proceedings of the 2011 IEEE 52Nd Annual Symposium on Foundations of Computer Science.* FOCS '11, Washington, DC, USA, IEEE Computer Society (2011), 97–106.

[44] Z. Brakerski, C. Gentry, and V. Vaikuntanathan, (leveled) fully homomorphic encryption without bootstrapping. In: *Proceedings of the 3rd Innovations in Theoretical Computer Science Conference.* ITCS '12, New York, NY, USA, ACM (2012), 309–325.

[45] M. van Dijk, C. Gentry, S. Halevi, and V. Vaikuntanathan, Fully homomorphic encryption over the integers. In: *Proceedings of the 29th Annual International Conference on Theory and Applications of Cryptographic Techniques.* EUROCRYPT'10, Berlin, Heidelberg, Springer-Verlag (2010), 24–43.

[46] J. Katz and Y. Lindell, *Introduction to modern cryptography.* CRC press (2014).

[47] V. Danos and E. Kashefi, Determinism in the one-way model. *Physical Review A* **74** (5) (2006), 052310.

[48] S. Aaronson, A. Cojocaru, A. Gheorghiu, and E. Kashefi, On the implausibility of classical client blind quantum computing. arXiv:1704.08482 (2017).

[49] V. Dunjko and E. Kashefi, Blind quantum computing with two almost identical states. arXiv:1604.01586 (2016).

[50] V. Dunjko, J.F. Fitzsimons, C. Portmann, and R. Renner, Composable security of delegated quantum computation. In: *International Conference on the Theory and Application of Cryptology and Information Security,* Springer (2014), 406–425.

[51] E. Kashefi and P. Wallden, Garbled quantum computation. *Cryptography* **1** (1) (2017), 6.

[52] T. Kapourniotis, V. Dunjko, and E. Kashefi, On optimising quantum communication in verifiable quantum computing. arXiv:1506.06943 (2015).

[53] H. Barnum, C. Crépeau, D. Gottesman, A.D. Smith, and A. Tapp, Authentication of quantum messages. In: *43rd Symposium on Foundations of Computer Science*

(FOCS 2002), 16-19 November 2002, Vancouver, BC, Canada, Proceedings (2002), 449–458.

[54] D. Aharonov and M. Ben-Or, Fault-tolerant quantum computation with constant error rate. *SIAM Journal on Computing* **38** (4) (2008), 1207–1282.

[55] D. Gottesman and I.L. Chuang, Demonstrating the viability of universal quantum computation using teleportation and single-qubit operations. *Nature* **402** (6760) (Nov 1999), 390–393.

[56] E. Kashefi and P. Wallden, Optimised resource construction for verifiable quantum computation. *Journal of Physics A: Mathematical and Theoretical* **50** (14) (2017), 145306.

[57] R. Raussendorf, J. Harrington, and K. Goyal, A fault-tolerant one-way quantum computer. *Annals of physics* **321** (9) (2006), 2242–2270.

[58] R. Raussendorf, J. Harrington, and K. Goyal, Topological fault-tolerance in cluster state quantum computation. *New Journal of Physics* **9** (6) (2007), 199.

[59] K. Fisher, A. Broadbent, L. Shalm, Z. Yan, J. Lavoie, R. Prevedel, T. Jennewein, and K. Resch, Quantum computing on encrypted data. *Nature communications* **5** (2014), 3074.

[60] C. Crépeau, Cut-and-choose protocol. In: *Encyclopedia of Cryptography and Security.* Springer (2011), 290–291.

[61] E. Kashefi, L. Music, and P. Wallden, The quantum cut-and-choose technique and quantum two-party computation. arXiv:1703.03754 (2017).

[62] T. Morimae, D. Nagaj, and N. Schuch, Quantum proofs can be verified using only single-qubit measurements. *Physical Review A* **93** (2) (2016), 022326.

[63] A.Y. Kitaev, A. Shen, and M.N. Vyalyi, *Classical and quantum computation.* Volume 47. American Mathematical Society, Providence (2002).

[64] J. Kempe, A. Kitaev, and O. Regev, The complexity of the local hamiltonian problem. *SIAM Journal on Computing* **35** (5) (2006), 1070–1097.

[65] J. Bausch and E. Crosson, Increasing the quantum unsat penalty of the circuit-to-hamiltonian construction. arXiv:1609.08571 (2016).

[66] D. Mayers and A. Yao, Self testing quantum apparatus. *Quantum Info. Comput.* **4** (4) (July 2004), 273–286.

[67] A. Coladangelo and J. Stark, Separation of finite and infinite-dimensional quantum correlations, with infinite question or answer sets. arXiv:1708.06522 (2017).

[68] B. Cirel'son, Quantum generalizations of bell's inequality. *Letters in Mathematical Physics* **4** (2) (1980), 93–100.

[69] J.F. Clauser, M.A. Horne, A. Shimony, and R.A. Holt, Proposed experiment to test local hidden-variable theories. *Phys. Rev. Lett.* **23** (Oct 1969), 880–884.

[70] M. McKague, T.H. Yang, and V. Scarani, Robust self-testing of the singlet. *Journal of Physics A: Mathematical and Theoretical* **45** (45) (2012), 455304.

[71] H.L. Huang, Q. Zhao, X. Ma, C. Liu, Z.E. Su, X.L. Wang, L. Li, N.L. Liu, B.C. Sanders, C.Y. Lu, et al., Experimental blind quantum computing for a classical client. *Physical Review Letters* **119** (5) (2017), 050503.

[72] J. Barrett, L. Hardy, and A. Kent, A., No signaling and quantum key distribution. *Physical review letters* **95** (1) (2005), 010503.

[73] A. Acín, N. Brunner, N. Gisin, S. Massar, S. Pironio, and V. Scarani, Device-independent security of quantum cryptography against collective attacks. *Physical Review Letters* **98** (23) (2007), 230501.

[74] E. Schrödinger, Probability relations between separated systems. *Mathematical Proceedings of the Cambridge Philosophical Society* **32** (10) (1936), 446–452.

[75] J. Fitzsimons and T. Vidick, A multiprover interactive proof system for the local hamiltonian problem. In: *Proceedings of the 2015 Conference on Innovations in Theoretical Computer Science*, ACM (2015), 103–112.

[76] R. Laflamme, C. Miquel, J.P. Paz, and W.H. Zurek, Perfect quantum error correcting code. *Physical Review Letters* **77** (1) (1996), 198.

[77] Z. Ji, Classical verification of quantum proofs. In: *Proceedings of the forty-eighth annual ACM symposium on Theory of Computing*, ACM (2016), 885–898.

[78] N.D. Mermin, Simple unified form for the major no-hidden-variables theorems. *Physical Review Letters* **65** (27) (1990), 3373.

[79] A. Peres, Incompatible results of quantum measurements. *Physics Letters A* **151** (3–4) (1990), 107–108.

[80] E. Knill and R. Laflamme, Power of one bit of quantum information. *Physical Review Letters* **81** (25) (1998), 5672.

[81] T. Kapourniotis, E. Kashefi, and A. Datta, Verified delegated quantum computing with one pure qubit. arXiv:1403.1438 (2014).

[82] M.J. Bremner, R. Jozsa, and D.J. Shepherd, Classical simulation of commuting quantum computations implies collapse of the polynomial hierarchy. In: *Proceedings of the Royal Society of London A: Mathematical, Physical and Engineering Sciences*, The Royal Society (2010), rspa20100301.

[83] D. Mills, A. Pappa, T. Kapourniotis, and E. Kashefi, Information theoretically secure hypothesis test for temporally unstructured quantum computation. arXiv:1704.01998 (2017).

[84] T. Kapourniotis and A. Datta, Nonadaptive fault-tolerant verification of quantum supremacy with noise. arXiv:1703.09568 (2017).

[85] E. Ising, Beitrag zur theorie des ferromagnetismus. *Zeitschrift für Physik A: Hadrons and Nuclei* **31** (1) (1925), 253–258.

[86] X. Gao, S.T. Wang, and L.M. Duan, Quantum supremacy for simulating a translation-invariant ising spin model. *Physical Review Letters* **118** (4) (2017), 040502.

[87] L. Disilvestro and D. Markham, Quantum protocols within spekkens' toy model. *Physical Review A* **95** (5) (2017), 052324.

[88] R.W. Spekkens, Evidence for the epistemic view of quantum states: A toy theory. *Physical Review A* **75** (3) (2007), 032110.

[89] M.A. Nielsen and I.L. Chuang, *Quantum Computation and Quantum Information: 10th Anniversary Edition*. 10th edn. Cambridge University Press, New York, NY, USA (2011).

[90] H. Buhrman, R. Cleve, M. Laurent, N. Linden, A. Schrijver, and F. Unger, New limits on fault-tolerant quantum computation. In: *Foundations of Computer Science, 2006. FOCS'06. 47th Annual IEEE Symposium*, IEEE (2006), 411–419.

[91] K. Fujii and M. Hayashi, Verifiable fault-tolerance in measurement-based quantum computation. arXiv:1610.05216 (2016).

[92] S. Barz, J.F. Fitzsimons, E. Kashefi, and P. Walther, Experimental verification of quantum computation. *Nat. Phys.* **9** (11) (Nov 2013), 727–731 article.

[93] S. Barz, E. Kashefi, A. Broadbent, J.F. Fitzsimons, A. Zeilinger, and P. Walther, Demonstration of blind quantum computing. *Science* **335** (6066) (2012), 303–308.

[94] C. Greganti, M.C., Roehsner, S. Barz, T. Morimae, and P. Walther, Demonstration of measurement-only blind quantum computing. *New Journal of Physics* **18** (1) (2016), 013020.

[95] C. Greganti, M.C. Roehsner, S. Barz, M. Waegell, and P. Walther, Practical and efficient experimental characterization of multiqubit stabilizer states. *Physical Review A* **91** (2) (2015), 022325.

[96] Ibm quantum experience. `http://research.ibm.com/ibm-q/`.

[97] Ibm 16-qubit processor. `https://developer.ibm.com/dwblog/2017/quantum-computing-16-qubit-processor/`.

[98] Google 49-qubit chip. `https://spectrum.ieee.org/computing/hardware/google-plans-to-demonstrate-the-supremacy-of-quantum-computing`.

[99] A. Broadbent and S. Jeffery, Quantum homomorphic encryption for circuits of low T-gate complexity. In: *Advances in Cryptology – CRYPTO 2015* – 35th Annual Cryptology Conference, Santa Barbara, CA, USA, August 16–20, 2015, Proceedings, Part II (2015), 609–629.

[100] Y. Dulek, C. Schaffner, and F. Speelman, Quantum Homomorphic Encryption for Polynomial-Sized Circuits. In: *Advances in Cryptology – CRYPTO 2016*, Springer, Berlin, Heidelberg (2016), 3–32.

[101] G. Alagic, Y. Dulek, C. Schaffner, and F. Speelman, Quantum fully homomorphic encryption with verification. arXiv:1708.09156 (2017).

[102] U. Mahadev, Classical homomorphic encryption for quantum circuits. arXiv:1708.02130 (2017).

[103] A. Shamir, Ip= pspace. *Journal of the ACM* (JACM) **39** (4) (1992), 869–877.

[104] D. Aharonov, I. Arad, and T. Vidick, Guest column: the quantum pcp conjecture. *Sigact News* **44** (2) (2013), 47–79.

[105] J. Watrous, Guest column: An introduction to quantum information and quantum circuits 1. *Sigact News* **42** (2) (June 2011), 52–67.

[106] J. Watrous, Quantum computational complexity. In: *Encyclopedia of complexity and systems science.* Springer (2009), 7174–7201.

[107] N. Harrigan and R.W. Spekkens, Einstein, incompleteness, and the epistemic view of quantum states. *Foundations of Physics* **40** (2) (2010), 125–157.

[108] D. Gottesman, An introduction to quantum error correction and fault-tolerant quantum computation. In: *Quantum information science and its contributions to mathematics*, Proceedings of Symposia in Applied Mathematics. Volume 68 (2009), 13–58.

[109] R. Raussendorf and H.J. Briegel, A one-way quantum computer. *Phys. Rev. Lett.* **86** (May 2001), 5188–5191.

[110] H.J. Briegel, D.E. Browne, W. Dur, R. Raussendorf, and M. Van den Nest, Measurement-based quantum computation. *Nat. Phys.* (Jan 2009), 19–26.

[111] Complexity Zoo. `https://complexityzoo.uwaterloo.ca/Complexity_Zoo`.

[112] S. Arora and B. Barak, *Computational Complexity: A Modern Approach.* 1st edn. Cambridge University Press, New York, NY, USA (2009).

[113] M. Ben-Or, S. Goldwasser, J. Kilian, and A. Wigderson, Multi-prover interactive proofs: How to remove intractability assumptions. In: *Proceedings of the twentieth annual ACM symposium on Theory of computing*, ACM (1988), 113–131.

[114] R. Cleve, P. Hoyer, B. Toner, and J. Watrous, Consequences and limits of nonlocal strategies. In: *Computational Complexity*, 2004. Proceedings. 19th IEEE Annual Conference on, IEEE (2004), 236–249.

Alexandru Gheorghiu
Institute for Theoretical Studies, ETH Zürich
Switzerland
e-mail: `agheorghiu@ethz.ch`

Theodoros Kapourniotis
Department of Physics, University of Warwick
UK
e-mail: `T.Kapourniotis@warwick.ac.uk`

Elham Kashefi
School of Informatics, University of Edinburgh
UK
and
CNRS LIP6, Université Pierre et Marie Curie, Paris
France
e-mail: `ekashefi@gmail.com`

The manufacturer's authorised representative in the EU is Springer Nature Customer Service Centre GmbH, Europaplatz 3, 69115 Heidelberg, Germany. If you have any concerns regarding our products, please contact ProductSafety@springernature.com

Printed and bound by CPI Group (UK) Ltd, Croydon, CR0 4YY
05/08/2026
02187214-0001